A HANDBOOK FOR CHARTERED SURVEYORS

PROFESSIONAL CONDUCT

A HANDBOOK FOR CHARTERED SURVEYORS

PROFESSIONAL CONDUCT

Richard Chalkley

R.I.C.S. BOOKS

Published by Surveyors Holdings Limited,
a wholly owned subsidiary of
The Royal Institution of Chartered Surveyors,
under the RICS Books imprint
12 Great George Street
London SW1P 3AD

A CIP catalogue record for this title is available from the British Library.

ISBN 0 85406 477 X

Designed by Dale Dawson

Design and production in association with
Book Production Consultants, 47 Norfolk Street, Cambridge

Typeset by Cambridge Photosetting Services

Printed in England by St. Edmundsbury Press,
Bury St. Edmunds, Suffolk.

AUTHOR'S NOTE

References to 'Bye-Laws', 'Regulations' and 'Rules' are to the Bye-Laws and Regulations of The Royal Institution of Chartered Surveyors.

I have also reproduced relevant proposed Bye-Law and Regulation changes which were approved by the Extraordinary General Meeting (EGM) of the Institution in October 1990. They will not come into effect until Her Majesty's Privy Council have given approval to them. The date of their coming into effect will be publicised in *Chartered Surveyor Weekly*, but the reader should contact either the Professional Conduct Department or the Standards and Practice Department of the Institution (telephone 071 222 7000) for current information as to the date of their taking effect.

The Institution constantly reviews its Bye-Laws and Regulations and in recent years there have been some quite substantial alterations (for example Members' Accounts Regulations, Bye-Law (24)(9A) and the Compulsory Professional Indemnity Insurance Regulations). While future editions of this book will incorporate any changes and modifications to the Institution's Rules, the reader is cautioned to check with the Professional Conduct Department and make sure there have been no alterations.

Words expressed in the masculine include, where appropriate, the feminine.

FOREWORD

Professions are largely a creature of public demand. They remain in existence because of the continuing recourse by the public to them; and they can only survive if the public has confidence in them.

The fact that the professions can command public confidence rests on two essential elements, professional knowledge and ethical conduct.

This guide is aimed at some areas of conduct and is written in a simple direct way to illustrate how the law and RICS Bye-Laws complement and overlap each other. It will be of interest to all Chartered Surveyors and TPC candidates. Richard Chalkley is to be congratulated for conceiving a guide in this form and for writing it.

Ian V. Oddy, FRICS, IRRV, FCIArb.

Chairman, RICS Standards and Practice Committee

PREFACE

This book is written primarily for newly qualified chartered surveyors and those about to set up in practice. However, I hope that some of the more experienced members of the profession and those who are not engaged in private practice will find it useful too. While it is only intended to be a guide, I have attempted to explain some of the most frequently encountered aspects of company and partnership law as it applies to practising surveyors, and also the relevant parts of the Bye-Laws and Regulations of The Royal Institution of Chartered Surveyors.

The views expressed are mine and do not bind The Royal Institution of Chartered Surveyors.

ACKNOWLEDGEMENTS

In producing this book I have been helped and encouraged by many people. My thanks are extended to the Directors of Surveyors Holdings Limited for their kind permission to reproduce extracts from the Institution's Bye-Laws and Regulations; to the Directors of RICS Insurance Services Limited for permission to reproduce the Professional Indemnity Collective Policy; to the Directors of Sweet and Maxwell Limited and Magna Legal Precedents Limited for permission to reproduce the sample precedents; to the Institution's Librarian, Pauline Lane-Gilbert, for compiling a superb index and to my secretaries, Paula Broad and Norma Coffey, for having typed and proof-read the manuscript. I am also greatly indebted to the staff of the Professional Conduct Department of the RICS, both present and past, for their invaluable guidance, help and encouragement; and to those chartered surveyors who were kind enough to read my manuscript and suggest improvements, not least to Ian Oddy, FRICS IRRV FCIArb, Chairman of the Standards and Practice Committee and past Chairman of the Professional Practice Committee of the RICS who also wrote the Foreword. Any errors occurring in the text are mine.

Richard Chalkley
Invicta House
Maidstone
Summer 1990

CONTENTS

APPENDICES

Chapter One

INTRODUCTION TO THE INSTITUTION'S PROFESSIONAL RULES

> 'A profession involves a particular kind of relationship with clients, or patients, arising from the complexity of the subject matter which deprives the client of the ability to make informed judgments for himself and so renders him to a large extent dependent upon the professional man. A self imposed code of professional ethics is intended to correct the imbalance in the relationship between the professional man and his client and resolve the inevitable conflicts between the interests of the client and the professional man or of the community at large.'

So said the Ormrod Committee Report on Legal Education, 1971 (Cmnd. 4595, P. 35 para. 86).

A professional code is the judgement of a profession on how its members should conduct themselves. The primary purpose of any code is to ensure the protection of the public. It is the responsibility of each member of a profession to uphold his or her professional code; failure to do so damages the reputation of the whole profession.

In their professional roles – both as employer and employee – chartered surveyors can probably do most to maintain the high standards of their profession by providing their services efficiently, competently and quickly at a fee that fairly reflects the amount of time, effort and expertise called for.

Surveyors practise in an increasingly competitive environment.

Some of the functions performed by chartered surveyors (for example auctioneering, estate agency and property management) have been performed by others for years. We are now seeing added competition in the market place from solicitors, accountants, architects, and other professionals. Chartered surveyors have already learned the need to be responsive to changing client attitudes and the Institution's professional rules have undergone substantial changes in the past few years to provide a framework which enables Members to compete on an equal footing while, at the same time, protecting the public.

There are several ways in which a chartered surveyor might practise and different obligations and responsibilities attach to each of them. I have sought in this guide to go through the five main ways and to explain not only their different legal characteristics, but also how the Institution's Rules distinguish between them. For illustrative purposes only I have invented two main characters, Ian M. Hopeful and Ann M. Bright and some subsidiary ones, whom I use to illustrate the problems practitioners can find themselves in. All the characters and the practices I mention are figments of my imagination.

The five main ways in which surveyors practise are as:

- employees;
- partners;
- directors;
- sole principals; and
- consultants.

While the Institution's Bye-Laws and Regulations apply to all chartered surveyors they will not all be relevant to all Members all the time since some of them are designed to deal with specific situations. For ease of reference I have reproduced the extracts from the Bye-Laws and Regulations which form the professional rules in Appendices I, II, III, IV, V, VI, XII, XIII and XIV.

In this guide I shall explain these provisions and draw Members' attention to those Rules which are relevant to them when they practise in the capacities set out above.

1. Bye-Laws and Regulations

First, I should explain the distinction between Bye-Laws and Regulations, since it may have already confused some of you. Bye-Laws normally express a principle which is considered to be of fundamental importance to the profession as a whole. They do not normally concern themselves with detail. The principles are likely to need modifying less often than the detail, and any change requires exposure to the whole membership at an Extraordinary General Meeting (EGM), and approval by at least two-thirds of those Members who vote. Before a change thus approved can come into effect, it must be allowed by Her Majesty's Privy Council – the guardians of the Institution's Royal Charter.

Regulations on the other hand normally contain matters of detail. Regulations may not impose obligations or grant rights beyond those contained in the principles of the Bye-Laws, but they may 'put flesh on the bones' of those principles. The General Council is empowered to make or change Regulations without an EGM, but it must first consult the Divisional Councils.

The only professional conduct Rules which are relevant to *all* Members *all* the time are those concerned with:

- 'Conduct Unbefitting' – Bye-Law 24(1);
- your chartered designation – Bye-Law 5;
- keeping in touch with the Institution – Bye-Law 24(9B);
- publicity – Regulations 16 and 17;
- keeping yourself up to date – Bye-Law 9(3); and the
- Continuing Professional Development Regulations.

2. 'Conduct Unbefitting': Bye-Law 24(1)

'No Member shall conduct himself in a manner unbefitting a Chartered Surveyor.'

This Bye-Law expresses the constant, continuing requirement that all chartered surveyors must conduct themselves properly at all times and in a way in which the public would expect professional people to behave. Unlike some other professions, the Institution is not governed by statute, but by its Royal Charter as amended from time to time. It is a self regulating profession. Like all professions, the standard of conduct expected of its Members must adapt to changing circumstances. Examples of cases in the past where Bye-Law 24(1) has been invoked include:

- a chartered surveyor's failure to pay monies due under a High Court judgment;
- the misappropriation by a chartered surveyor of his employer's money;
- the writing of a rude and offensive letter to a former client;
- a breach of client confidentiality;
- failure or delay in replying to correspondence from a client;
- failure or delay in replying to correspondence from another professional; and
- failure or delay in replying to correspondence from the Institution in a disciplinary enquiry.

Any breach of the Institution's other Bye-Laws or Regulations may itself also amount to a breach of Bye-Law 24(1). For example, a serious breach of client confidentiality, while infringing Conduct Regulation 19 (see Appendix III) could also amount to 'conduct unbefitting' in breach of Bye-Law 24(1).

3. Your chartered designation: Bye-Law 5

The full text of this Bye-Law is set out in Appendix II. It regulates the use by Members of the chartered designation and the distinguishing initials 'FRICS' and 'ARICS'. Your entitlement to describe yourself as a 'Chartered Surveyor' is one of the main privileges conferred on you by your membership of a professional body governed by Royal Charter. You have a duty to ensure its correct use, both by you and by others. I shall comment further upon this Bye-Law later in relation to partners and directors and partnerships and companies.

4. Keeping in touch with the Institution: Bye-Law 24(9B)

'Every Member shall, in accordance with the Regulations, furnish to the Institution such particulars of his practice, employment and business as it may reasonably require for the administration of the Institution and for the regulation of Members' professional conduct and discipline.'

This is a new requirement which became effective in January 1990. Under a Regulation which came into effect at the same time (Conduct Regulation 24), Members will periodically be required to supply the Institution, within 28 days of the request, with particulars so that Membership records can be regularly updated with details of employers, practice addresses, fields of practice and the like. Having once submitted the required details, Members must notify the Institution of any change *within seven days* of its taking place. These requirements are aimed at making communication between the Institution and its Members more accurate and therefore more cost effective, in Members' own interests.

5. Publicity: Yours – Regulation 16

'Every Member shall ensure that any publicity for which he may be held responsible is neither inaccurate nor misleading nor likely to cause public offence.'

This Regulation is drawn very widely. The publicity to which it refers would include not only all advertisements, auction catalogues and property sales particulars issued by a Member or his practice, but also newspaper articles, public talks – in fact any statement in the public domain, whether verbal, pictorial or cinematic – for which the Member could be held responsible. Take care! Not only must you avoid misleading or offending others yourself, but you must avoid allowing your employees or associates to do so.

6. Publicising the Institution: Regulation 17

'No Member shall:

(a) purport to represent the views of the Institution unless expressly authorised so to do; or

(b) publicise the Institution or its Members generally in terminology which has not either already appeared in an advertisement published by the Institution or received the approval of the Institution.'

The purpose of this Regulation is self evident and needs no amplification.

7. Keeping yourself up to date: Bye-Law 9(3) and the Continuing Professional Development Regulations

Bye-Law 9(3)

'Every Professional Associate and Fellow of the Institution shall for so long as he remains a Member undergo in each year such continuing professional development and shall from time to time provide to the Institution such evidence that he has done so as the Regulations shall provide.'

The Continuing Professional Development (CPD) Regulations are set out in Appendix IV.

Long gone are the days when a professional person could breathe a sigh of relief after passing his final examinations and throw his text books and lecture notes away. Professionalism requires a commitment to gain and use new knowledge and skills, particularly in today's fast changing world. Until 1991 the requirement on Members to undergo 'Continuing Professional Development' extends only to those elected corporate Members on or after 1st January 1981. The requirement will extend to all Fellows and Professional Associates, whenever elected, as from January 1991. The Institution has published an extremely helpful guide entitled 'Guide to Continuing Professional Development' which is available free of charge to Members from the Education and Membership Department of the Institution.

Chapter Two

CHARTERED SURVEYORS IN PRIVATE PRACTICE

8. Employees

After qualifying most surveyors will choose to be employees, either in private practice, industry, central or local government or in some other public authority. One of the first things to note is that the law and the Institution's Rules place heavier burdens on principals (i.e. partners, directors or sole practitioners) than on employees, and that employees must, therefore, be careful not to allow themselves to be 'held out' or represented to the public as principals. If employees lead others to believe that they are principals they will find that the law and the Institution treat them as having responsibility for the conduct of their employer's business. Since they are unlikely to have any authority over it, this is clearly to be avoided!

I shall discuss the additional responsibilities falling on principals (usually employers) in Section 9. Let us for the moment concentrate on how employees should represent themselves, and on the duties that fall to them.

8.1 Employees of partnerships

A Holding out as a partner

Let me introduce you to Ian M. Hopeful. Since leaving University Ian has worked for a local firm of chartered surveyors, Carefree, Jolly & Happy. On qualifying his employers decided to add his name to the notepaper. It looks like Figure 1.

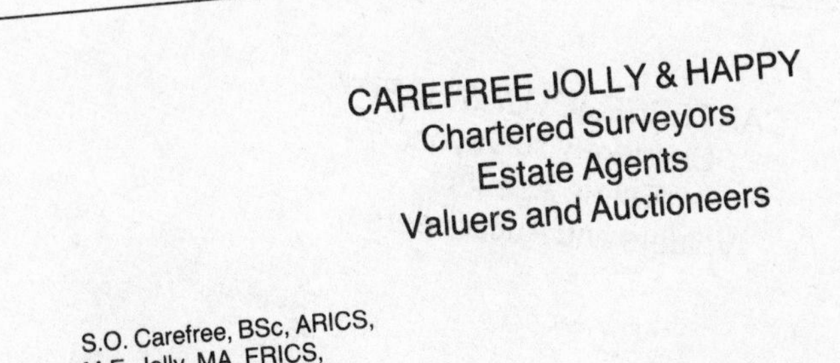
CAREFREE JOLLY & HAPPY
Chartered Surveyors
Estate Agents
Valuers and Auctioneers

S.O. Carefree, BSc, ARICS,
M.E. Jolly, MA, FRICS,
M.R. Happy, ARICS

Staff Surveyor:
I.M. Hopeful, BSc, ARICS

1 High Street,
Anytown.
Tel: Anytown 7000

Figure 1

Because he is expressly described as 'Staff Surveyor' it will be clear to anyone reading the notepaper that Ian is an employee and not a partner. The line drawn beneath the name 'M.R. Happy' would not by itself indicate clearly that Ian was not a partner. Were it not for the words 'Staff Surveyor' the addition of his name beneath the list of partners would imply that he was a partner. He would then become responsible both under the law and under the Institution's Bye-Laws and Regulations for the activities of the partnership in just the same way as the partners (see Section 9 'Partners'). Let us look at another example, Figure 2.

Here, the notepaper is not clear. It might be thought, by anyone reading it, that Ian and his colleagues, H.E. Sells and U. Care, are partners with specialist responsibilities. A much better way of showing that they are not partners, but senior staff, would be to set out the notepaper as in Figure 3.

The addition of the headings to the two columns leaves no one in any doubt as to who are the partners and, therefore, responsible for the activities of the business, and who are not.

Let me give a further example. Some practices use the description 'Associate' to describe senior personnel. In the

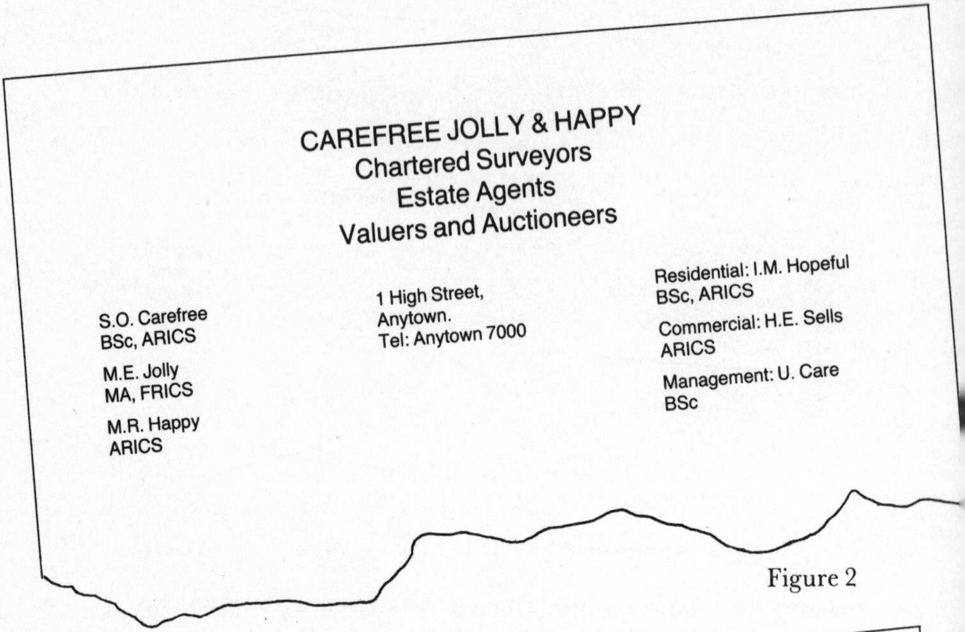

CAREFREE JOLLY & HAPPY
Chartered Surveyors
Estate Agents
Valuers and Auctioneers

S.O. Carefree
BSc, ARICS

M.E. Jolly
MA, FRICS

M.R. Happy
ARICS

1 High Street,
Anytown.
Tel: Anytown 7000

Residential: I.M. Hopeful
BSc, ARICS

Commercial: H.E. Sells
ARICS

Management: U. Care
BSc

Figure 2

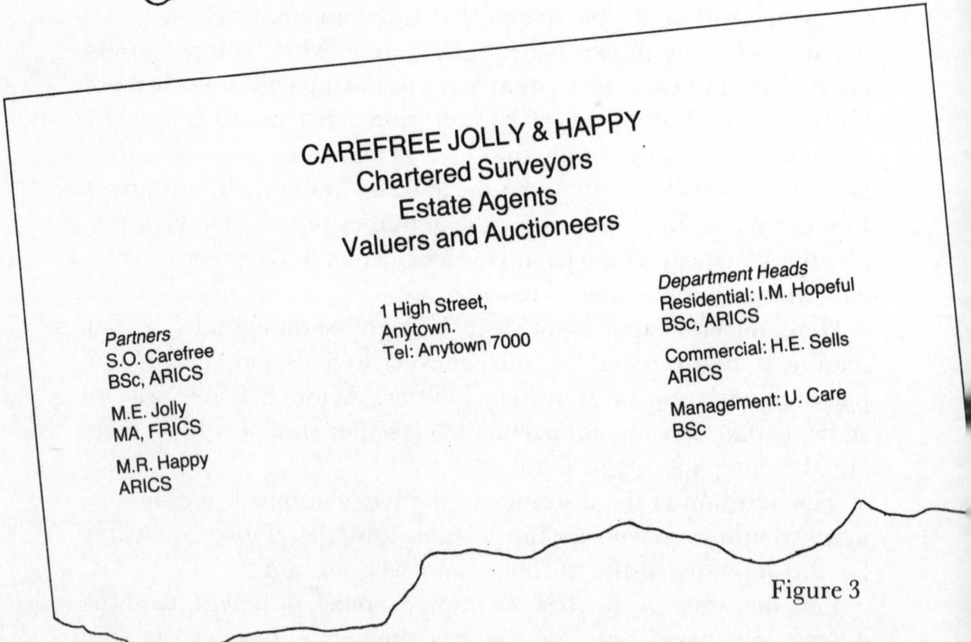

CAREFREE JOLLY & HAPPY
Chartered Surveyors
Estate Agents
Valuers and Auctioneers

Partners
S.O. Carefree
BSc, ARICS

M.E. Jolly
MA, FRICS

M.R. Happy
ARICS

1 High Street,
Anytown.
Tel: Anytown 7000

Department Heads
Residential: I.M. Hopeful
BSc, ARICS

Commercial: H.E. Sells
ARICS

Management: U. Care
BSc

Figure 3

following example Ian has been promoted further. His employers wish to recognise that fact publicly by describing him as an 'Associate' on the firm's notepaper. The letter heading might look like Figure 4.

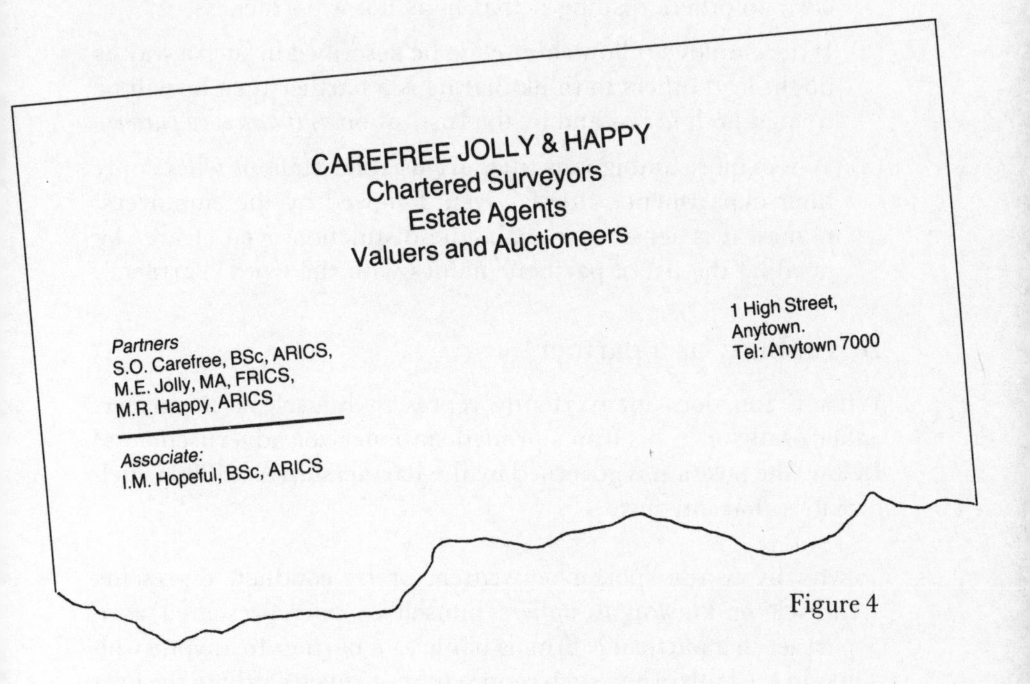

CAREFREE JOLLY & HAPPY
Chartered Surveyors
Estate Agents
Valuers and Auctioneers

1 High Street,
Anytown.
Tel: Anytown 7000

Partners
S.O. Carefree, BSc, ARICS,
M.E. Jolly, MA, FRICS,
M.R. Happy, ARICS

Associate:
I.M. Hopeful, BSc, ARICS

Figure 4

So far so good. The description 'Associate' does not in itself suggest that Ian is a partner.

But what would be the position if he were described as 'Associate Partner'?

This description clearly suggests that Ian *is* a partner of some sort. He therefore becomes responsible in law for the firm's activities even if in fact he is still an employee. Ian should avoid this situation at all costs. As an employee he should not allow himself to be described by any phrase which includes the word 'partner', such as 'Associate Partner' or 'Local Partner'. (If on the other hand Ian is in fact a salaried partner, then he is a partner in law and may represent himself as such (see Section B below).)

To sum up therefore:

(1) If an employee's name appears on the firm's notepaper or elsewhere (e.g. advertising literature, sales particulars, name-boards, visiting cards etc.) it must appear in such a way as to be clear to others reading it that he is not a partner.

(2) If the employee allows himself to be described in such a way as might lead others to think that he is a partner then he will be treated both in law and by the Institution *as if he were a partner*.

(3) Where more ambiguous titles are used for staff, or where only their department's title is given, followed by the employees' names, it is sensible to make the distinction even clearer by heading the list of partners' names with the word 'Partners'.

B Liability as a partner

What if Ian does inadvertently represent himself as a partner, either orally or in his firm's printed stationery or advertisements? In law, the position is governed by the Partnership Act 1890 which provides that anyone:

> 'who by words spoken or written, or by conduct, represents himself or knowingly suffers himself to be represented as a partner in a particular firm, is liable as a partner to anyone who has in the faith of any such representation given credit to the firm whether the representation has or has not been made or communicated to the person so giving credit by or with the knowledge of the apparent partner making the representation or suffering it to be made.'

If, therefore, credit is given to a partnership based upon representations made by someone who is purporting to be a partner, but who in fact is not a partner, then that person is equally responsible to the person giving credit along with the partners. He is fully liable to the party misled. Partners also have many onerous legal responsibilities (see Section 9). This is why anyone who is not a partner should be very careful in all his dealings with third parties not to imply to others that he is a partner, either on paper or orally.

The Institution would similarly place on Ian the responsibilities that it places on partners in relation to Compulsory Professional Indemnity Insurance (see Section 9.2A), the Members' Accounts Regulations (see Section 9.2B) and vicarious liability for the conduct of others in the firm (see Section 9.2I).

C Non member firms: Professional Indemnity Insurance and vicarious liability

There is one situation in which Ian would be required by the Institution *as an employee* to take responsibility for his firm's professional indemnity insurance. That is if Ian's firm had no chartered surveyor partners *and* Ian's name or designatory letters or chartered designation appeared on the firm's stationery, however clearly distinguished from the partners. (See definition of 'Consultant' in Regulation 1 of the Compulsory Professional Indemnity Insurance Regulations, Appendix V.)

In this situation, Ian would have the unenviable task of persuading his non Member employers to effect professional indemnity insurance that complies with the Institution's requirements, and to certify to the Institution that they have done so, or face possible disciplinary proceedings.

Where Ian's employers are not chartered surveyors, Ian also has a heavier duty to the Institution to see that its other Rules are complied with throughout his employers' practice if his name and/or designatory letters or designation appear on the notepaper or in the advertisements of the firm (Bye-Law 24(5)(b)(ii) – see Appendix I).

Unless he can show that he could not have known of the contravention and that he took reasonable steps to prevent such an occurence, Ian may find himself penalised in disciplinary proceedings.

The purpose of these apparently burdensome responsibilities is to protect clients who may have chosen Ian's firm because it boasts a connection with the Institution through Ian's name and/or qualifications, which appear on stationery or in advertisements.

What should Ian do if he cannot persuade his employers to comply? A useful precaution, before accepting employment in a non-Member firm, is to obtain a written undertaking from the partners that they will comply with the Institution's Rules for so long as Ian is employed by them. Such an undertaking could go some way to help Ian show that he had taken reasonable steps to prevent a contravention of the Institition's Rules in any disciplinary action in which he might become involved because of the conduct of his non-Member colleagues.

Alternatively, Ian could try to ensure that neither his name, his designation nor his designatory letters ever appear on his firm's stationery or in advertisements. As a last resort, Ian might consider it wiser to leave the firm rather than jeopardise his own professional status and the reputation of the profession generally.

These may appear stark choices early in Ian's career, but he has worked hard to become a Member of a distinguished profession and that brings responsibilities, as well as privileges.

D Use of chartered designation

The Institution's Rules place another direct duty on Ian as an employee in private practice, if his name appears in any list of either staff or partners, whether or not there are other chartered surveyors in the firm.

Regulation 13 provides:

'In any list of partners and/or staff in a firm, or directors and/or staff of a company, carrying on practice as surveyors, published by or on behalf of a Member or such firm or company, which includes the names of one or more Members, it shall be the duty of every Member or Members so named to ensure that his or their status within the firm or company is clearly stated and that no chartered designation is used in such a way as to give the impression that the firm or company is entitled to use that designation if that is not the case.'

This goes beyond the duty of all chartered surveyors to ensure that they use the chartered designation properly (that is, as permitted by Bye-Law 5) in relation to their own name. Once Ian's name appears in printed form on any list, he has a duty to ensure that the chartered designation is not improperly used in relation to his firm. I will be explaining the proper use of the chartered designations in relation to firms in Section 9.4 and in relation to companies in Section 10.4.

E When is an employee not an employee?

In the preceding paragraphs I have covered Ian's main professional responsibilities as an employee, including those that would fall to him if he held himself out as an employer.

I must also briefly draw attention to what can be a pitfall for some employees: that is when they act in their own right for people who are not clients of their employer's practice. I have in mind situations like these:

(a) While Ian is employed by Carefree, Jolly & Happy, he agrees to go one Saturday to look at a property a friend is thinking of buying, 'just to make sure there is nothing wrong with it'.

(b) As a member of his Church Restoration Committee, Ian is asked to obtain planning consent for an extension to the Church Hall.

(c) Ian's sister and brother-in-law have found cracks in their kitchen. They invite him for lunch one Sunday and ask him to advise them on what they should do about the cracks.

In none of these situations is Ian acting for his employer. He may not realise it, but he is in fact acting as a 'sole principal' (see Section 11). He will be liable in law for the advice he gives and is required by the Institution's Rules to be covered for professional indemnity insurance. It is unlikely that his employers' policy would insure him against risks he undertakes in his own right. Ian faces the same risks whether or not he is paid for his advice and whether or not it is in writing. He should either take out his own policy or, if possible, ensure that he is covered for these activities by an extension to his employers' policy.

8.2 Employees of incorporated practices

A Holding out as a director

These days many surveyors practise in companies rather than partnerships. Surveyors employed by companies must be just as careful as those employed by partnerships not to imply that they are principals. The legal basis is different from that in partnership law, but the consequences of an employee purporting to be an officer of the company (director) can be just as perilous.

In Company Law the term 'director' means anyone who is a member of the managing board of the company and *includes anyone occupying the position of a director by whatever name called* (Section 741 Companies Act 1985). A company is not required to display the names of its directors on its notepaper, but if it does show the name of any one of its directors then the names of all the directors must be shown.

Let me now introduce you to Ann M. Bright who has been working for a competitor practising in Anytown, Anytown Mega Estates Limited. She is the only chartered surveyor employed by the company. To give her the status her employers consider she deserves, she is shown on the notepaper as 'Survey Director'. The notepaper looks like Figure 5.

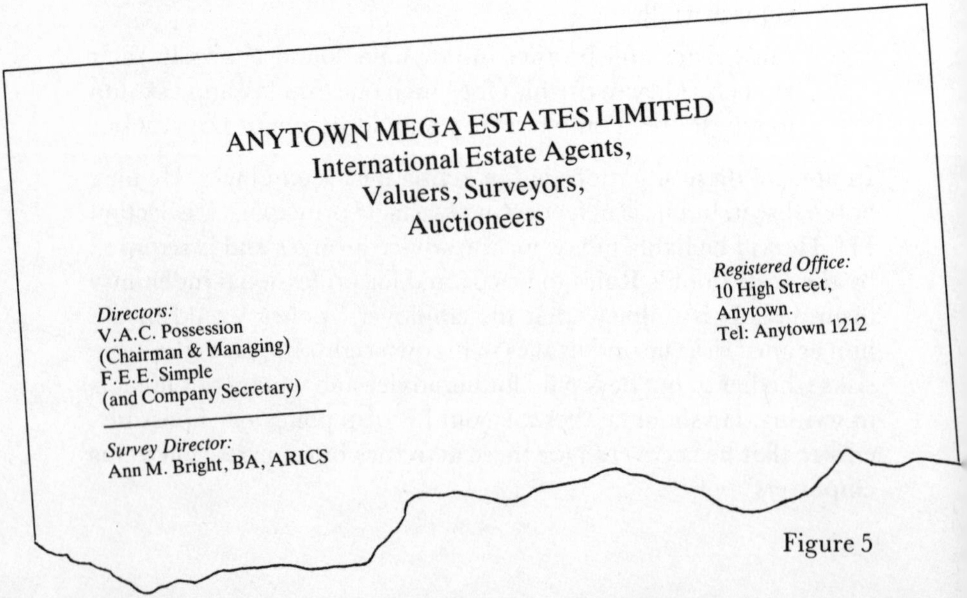

ANYTOWN MEGA ESTATES LIMITED
International Estate Agents,
Valuers, Surveyors,
Auctioneers

Directors:
V.A.C. Possession
(Chairman & Managing)
F.E.E. Simple
(and Company Secretary)

Survey Director:
Ann M. Bright, BA, ARICS

Registered Office:
10 High Street,
Anytown.
Tel: Anytown 1212

Figure 5

In the above example, Ann's name appears printed at the top of the notepaper beneath the names of the Chairman and his co-director, who is also the Company Secretary. Alternatively, she might merely be described as 'Survey Director' beneath her signature at the bottom of the notepaper. Her position in law would be the same. But what is it?

Is she a director? Does she assume the responsibilities and obligations of a director? Because her status is described ambiguously the answer is not clear cut. She may fall foul of company law; she almost certainly will be regarded by the Institution as having the responsibilities of a director (see Section 10). Does her title mean that she is a senior employee who has responsibility for surveys? Does it mean that she is the member of the Board of Directors of Anytown Mega Estates who has particular responsibility for surveys? If, without the knowledge and consent of the Board of Directors, she were to enter into a contract on behalf of the company, the company would be in some difficulty in subsequently denying the contract purely because Ann was not appointed to the Board and had not been given express authority. The same would be the case if she described herself as 'Sales Director', 'New Homes Director', 'Commercial Director' or used any other description in conjunction with the word 'Director'.

The description 'Director' should be avoided by persons not appointed to the Board of Directors of the company. 'Manager' or 'Executive' or 'Department Head' are less confusing descriptions to use and are, therefore, to be preferred.

B Other responsibilities

Ann's responsibilities to the Institution as an employee of a company are the same as Ian's as an employee of a partnership. If she misrepresents herself as a director, she also assumes the responsibilities discussed under 'Director' in Section 10. If she succeeds in avoiding such misrepresentation, but allows her name, designation or designatory letters to appear on the company's stationery or advertisements she still has duties as an employee:

- to ensure that her employers comply with the Institution's professional indemnity insurance requirements (if none of the directors is a chartered surveyor);
- to take reasonable steps to prevent breaches of the Institution's Rules generally throughout the company (if none of the directors is a chartered surveyor);
- to see that no chartered designation is improperly used in relation to her company (*whether or not* any of the directors are chartered surveyors).

These duties were more fully discussed in Section 8.1 above. Whether you are employed in a partnership or a company, you need to be familiar with them.

9. Partners

Ian Hopeful has done well. He has exceeded all his employers' targets, he is well liked by the partners, their staff and by clients and his employers are anxious not to lose his services. His firm has recognised that he is partnership material and in order to retain his services has decided that the time has come to reward his efforts by making him a salaried partner. The firm has indicated to him that after a respectable period as a salaried partner he will be promoted to full equity partnership.

The notepaper is altered so that his name appears with those of Carefree, Jolly and Happy (see Figure 6). Ian's duties and responsibilities have also altered – dramatically!

9.1 Partners' legal responsibilities

A Joint and several liability

Partners, whether salaried or equity are 'jointly and severally liable'. They can all be sued jointly and individually for breaches of all contracts entered into by the firm (or by any of the partners on behalf of the firm), for torts committed by the firm (or by any of its

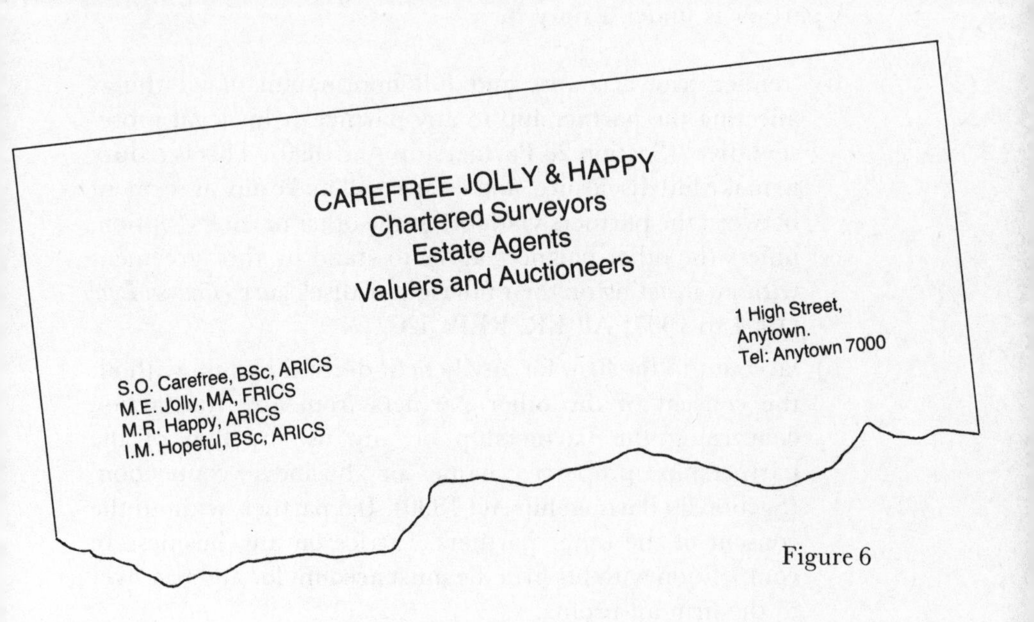

Figure 6

partners or employees) and they share all other liabilities (for example, taxation). This means that any one partner might end up paying for more than his or her 'fair' share of a claim. The Partnership Act 1890 allows partners to agree between themselves how their firm should operate, but where a partnership is not governed by agreement, or where there is a partnership agreement which does not exclude the operation of the Act, the provisions of the 1890 Act will apply. The Act provides that every partner is an agent of the firm and of his other partners. The decisions of each partner, taken in the usual course of the partnership business, bind both the partnership and all the partners, unless the individual in question has no authority to act for the firm in the matter, and the person with whom he is dealing either knows this, or doubts that he is a partner (Section 5 Partnership Act 1890). The extent of a partner's implied authority to bind the firm will depend on the nature of the firm's business, its normal practice and whether or not it is a trading partnership (see *Mercantile Credit Company Limited* v. *Garrod*, [1962] 3 All ER. 1103).

A partner is under a duty to:

(i) 'render true accounts and full information of all things affecting the partnership to any partner or his legal representative' (Section 28 Partnership Act 1890). This is a duty to make full disclosure, and failure will make any agreement between the partners voidable at the other partners' option, unless the other partners agree to stand by the agreement without insisting on their full right of disclosure (*Law* v. *Law* [1904 to 1907] All ER. REP. 526).

(ii) 'account to the firm for any benefit derived by him without the consent of the other partners from any transaction concerning the partnership or any use by him of the partnership property, name or business connection' (Section 29 Partnership Act 1890). If a partner, without the consent of the other partners, carries on any business in competition with his firm, he must account for and pay over to the firm all profits.

Ian Hopeful will be wise to enter into a partnership agreement to regulate the partnership. The provisions of the Act can be varied by agreement between the partners and it can be dangerous to rely solely on the Act without any agreement, especially in the case of salaried partners. Ian is, after all, only going to receive a salary in just the same way as the employees of the partnership. He will not share in the profits (or losses) as the equity partners will, yet *he will be liable as they are for all that goes on in the firm.*

B Salaried Partnership Agreements

I have set out at Appendix IX a suggested draft form of agreement for salaried partnerships. I would, however, add a word of caution. This suggested form *must not be used* without seeking professional legal, tax and accountancy advice. It is intended only to serve as a guide and is reproduced for illustrative purposes only. It is *not* intended to be a model.

It is particularly important in all salaried partnership agreements that the salaried partner, his or her personal representatives, estate and effects should be indemnified against all liabilities, claims, proceedings, costs, damages and expenses for which a partner would normally be responsible. It is also important that the salaried partners' powers should be restricted as provided for in Clause 5 of the draft Agreement.

A salaried partnership is very much a 'halfway house'. We have seen how Ian bears all the responsibilities of a partner, but has none of the rewards.

Although his liabilities can be covered to a limited extent by his partners' indemnity, an indemnity given by his partners is only as good as their ability to discharge those liabilities. Nor will an indemnity help him professionally. (See Section 9.2 below.)

C Equity Partnership Agreements

Ian is not always going to be a salaried partner. Eventually, he will be admitted to the partnership as a full equity sharing partner.

A simple partnership deed is reproduced at Appendix X. I must repeat the word of caution used previously. The sample agreement is not necessarily suitable for all partnerships and is intended only to serve as a check list. It *must not be used* without first seeking legal, tax and accountancy advice. It is *not* to be used as or regarded as being a model.

The agreement is a fairly straightforward one. It reflects that the business has been carried on for some time and spells out the type of business, the place of business and its name. I shall not dwell here on a full explanation of the Agreement, since many of the points it covers have been dealt with elsewhere in this book.

9.2 Partners' professional responsibilities

The following Bye-Laws and Regulations now become directly relevant to Ian as a partner in his own firm, whether salaried or equity.

A Compulsory Professional Indemnity Insurance: Bye-Law 24(9A) and the Compulsory Professional Indemnity Insurance Regulations

I have made reference elsewhere to these Regulations (see Sections 8.1 C, 8.2 B, and 11.2) which came into force on 1st January 1986. But what is professional indemnity insurance?

Although policies of insurance differ from one insurance company to another, essentially it is insurance cover providing the insured with an indemnity for claims made against him in the event of his being sued by clients or former clients for negligence, breach of contract, or other professional default (see Section 11.2).

Negligence is:

> 'the omission to do something which a reasonable man, guided upon those considerations which regulate the conduct of human affairs, would do, or something which a prudent and reasonable man would not do' (See *Blyth* v. *Birmingham Waterworks Co.* (1856) 11 Ex 781).

We are all under a duty to take reasonable care to avoid acts or omissions which we can reasonably foresee which would be likely to injure our neighbour. Lord Atkin defined 'neighbour' in the famous 'Ginger-beer' case, *Donoghue* v. *Stevenson* (1932) A.C. S.62 as:

> 'persons who are so closely and directly affected by any act that I ought reasonably to have them in contemplation as being so affected when I am directing my mind to the acts or omissions which are called in question'.

At Appendix XV I have reproduced the RICS Professional Indemnity Collective Policy. This is the policy referred to in the Compulsory Professional Indemnity Insurance Regulation 2(b). It provides wide cover because it indemnifies against *any* Civil liability, not just for liability arising out of negligence. All Members who are required to have indemnity insurance must have insurance which is no less comprehensive than this policy.

The Compulsory Professional Indemnity Insurance (CPII) Regulations represent one of the main 'pillars' of the Institution's Rules providing protection for the public. We are all human and make mistakes. The Regulations seek to ensure, in so far as is reasonably possible, that Members' clients do not suffer as a result.

But the Regulations provide considerable benefit to Members also. The costs of litigation today can be enormous. The mere bringing of an action by a client can impose considerable financial strain on Members, even if they are not found to have been negligent. The insurers will pay the Members' legal costs in defending actions, provided the Member complies with his obligations under the policy and notifies the insurers at the outset of the possibility of a claim.

Under the required policy, indemnity is also provided for:

(i) the Member's employees;
(ii) the Member's consultants; and for
(iii) the Member's former partners.

While Ian is an employee, the partners in Carefree, Jolly & Happy will be indemnified by its insurers for any civil liabilities arising out of his professional activities during the course of his employment, and so will he. The insurers will also provide indemnity to protect each partner against any dishonest or fraudulent acts or omissions by the staff or other partners, provided of course, that there was no collusion or connivance.

As an employee, Ian need not concern himself with indemnity insurance, (other than to seek confirmation from his employers that they are insured), but once he becomes a partner, *or is held out to be a partner*, he must arrange cover. In Ian's practice his partners need only to contact their insurers insurance brokers and advise them that he has joined the partnership as a partner.

Since insurance contracts are 'contracts of utmost good faith' Ian needs to ensure the *full* disclosure of any relevant facts concerning him, his area of practice, his size of practice, his partners and employees and of all claims made or threatened against him or the practice. If any relevant information is not disclosed then the insurers would be quite entitled to disclaim liability under the policy. It will not be until there is a claim and after the insurers have made enquiries that Ian's insurers disclaim liability. At that stage the consequences could be catastrophic for Ian.

In order to monitor compliance with Bye-Law 24(9A) and the CPII Regulations, the Institution requires Members to complete an annual CPII Certificate. This requires the Member to certify to the Institution whether or not he is subject to the Regulations and if he is, that he has complied with them.

The Institution does not normally ask for details of Member's insurance (although it has power to do so in Regulation 6) and it is, therefore, *for the Member to satisfy himself* that the insurance cover provided by his policy is adequate.

Insurance is not cheap and the scope of cover being offered varies widely from one company to another making it essential for Members to seek help from competent and reputable brokers, to ensure that his cover is no less comprehensive than the RICS Professional Indemnity Collective Policy issued by RICS Insurance Services Limited. A copy is available from either the Standards and Practice Department at the RICS or direct from RICS Insurance Services Limited, Plantation House, 31-35 Fenchurch Street, London EC3M 3DX (telephone 071 481 1445) and is reproduced at Appendix XV.

B Accounting for client's money: Bye-Law 24(8) and the Members' Accounts Regulations

As a partner, Ian is now directly responsible for all clients' money held by his firm. He becomes subject to the Members' Accounts Regulations, made under Bye-Law 24(8). The Bye-Law and the Regulations are reproduced in full at Appendices I and VI respectively. In particular, Ian is responsible for making returns to the Institution under Regulation 11(1). This imposes an obligation on Ian, once in each 12 month period, to provide the Institution with a certificate indicating whether he did or did not hold client's money. If neither he nor his firm did hold clients' money, Regulation 11(1)(a) provides that he must submit a 'nil' return within six months of the end of the accounting year in which no clients' money was held. If the partnership did hold clients' money then, of course, Ian is regarded as having held it since, as a partner, he is responsible for it. Under Regulation 11(1)(a) he must then submit a certificate to that effect, together with an Accountant's Report, no later than six months after the end of the accounting period to which it relates. The Members' Accounts Regulations apply to:

> *'every member who is a sole principal in a practice or a partner in a firm or a director of a company or is held out to the public as a partner in a firm or a director in a company carrying on practice as surveyors in the United Kingdom, the Isle of Man or the Channel Isles. . .'*

In a multi-partner practice one certficate may be submitted listing all the partners' names.

C Clients' money – what is it?

Let me first stress that I am only going to explain the meaning of 'clients' money' under the Institution's Members' Accounts Regulations. The term also has a statutory meaning under the Estate Agents Act 1979 and the Regulations made under that Act. The statutory meaning differs from what follows in several important ways. Estate agents' statutory obligations in respect of clients' money are also different from (though not inconsistent with) the obligations of chartered surveyors under the Institution's Rules. It is beyond the scope of this book to provide a commentary on the Estate Agents Act, but all chartered surveyors whose firms engage in estate agency *must* understand its provisions and comply with them. (There are several good guides available from Surveyors Bookshop, 12 Great George Street, Parliament Square, London SW1P 3AD, telephone 071 222 7000.)

Part I of the Members' Accounts Regulations contains the definitions. Regulation 1 defines 'clients' money' as meaning:

> *'money held or received by a Member, his firm or his company on account of a person for whom he, his firm or his company is acting either as a surveyor or, in connection with his practice as a surveyor, as agent, bailee or in any other capacity including that of a stakeholder provided that the expression 'clients' money' shall not include money to which the only person beneficially entitled is the Member himself, or money held in an account by a Member jointly with a third party (not being a client) and over which the Member does not have a power of withdrawal on his sole signature or of the signature of himself or any partner, co-director, servant, employee or other person connected with his firm.'*

'client' is defined as meaning *'any person or body on whose account a Member holds or receives clients' money'.*

To understand the definition of clients' money it is necessary to look at the definition in its constituent parts.

> *'money held or received by a Member, his firm or his company on account of a person for whom he, his firm or his company is acting either as a surveyor or, in connection with his practice as a surveyor, as agent, bailee or in any other capacity including that of stakeholder. . . '*

This clearly includes:

(i) rents;

(ii) building society valuation fees, where Ian is arranging a mortgage (as opposed to undertaking the valuation on behalf of the lending institution) on behalf of a client and receives the fee to pass on to the proposed lender;

(iii) service charges collected by Ian's firm when he is managing a property (but not his fees for acting as managing agent);

(iv) deposits and monies paid by a client on account of disbursements before they are actually incurred. (e.g. where monies are paid to Ian in respect of the fees due on an application for planning consent before those fees have been paid by Ian).

> *' . . . provided that the expression 'clients' money' shall not include money to which the only person beneficially entitled is the Member himself. . . '*

Ian cannot treat himself as a client and cannot therefore conduct his personal or office transactions through his client account, although, subject to the Regulations, he may put a 'float' in the client account from his own monies provided it is a nominal sum required to open the client account at a bank. Ian is not 'beneficially entitled' to money which he is holding as a Trustee for another person: this is clients' money and must be paid into a client account.

> *'. . . or money held in an account by a Member jointly with a third party (not being a client) and over which the Member does not have a power of withdrawal on his sole signature or the signature of himself or any partner, co-director, servant, employee or other person connected with his firm.'*

As one would expect, monies held in respect of service charges or representing sinking funds are always clients' monies. For example, monies held by a managing agent in a joint account with a landlord are also clients' monies where that landlord is a client. This is the case even if the Managing Agent needs the signature of his client or client's representative before he can withdraw monies. Monies not otherwise covered under the Regulation are not clients' monies. The occasions when a Member holds such money will be extremely few and far between. For monies on these grounds not to be clients' monies they must satisfy both conditions below. They must:

(a) be held jointly with a non client, *and*
(b) *not* be withdrawable on the Member's sole signature.

Doubt is often expressed as to how interest on clients' money should be accounted for. The Institution's Regulations do not cover this, but the law does. When chartered surveyors hold clients' money they are in a fiduciary relationship with their clients and are not entitled to make a profit from money entrusted to them. They can of course make proper charges to their clients for the professional services rendered, but without the client's agreement, a surveyor is not entitled to make indirect charges by way of retaining interest earned by the investment of clients' money.

However, provided the legal position is explained to the client and the client consents he may waive his right to interest. Any such waiver should be recorded in writing and should acknowledge the fact that the client understands that he is entitled to such interest. This is the position at law. There are also specific provisions covering the payment of interest in the Estate Agents Accounts Regulations made under the Estate Agents Act 1979.

Once again, let me remind you to study that Act, and the Regulations made under it, particularly in relation to clients' money, if you or your firm engage in estate agency.

D Clients' Money Protection Scheme

The Institution's Rules on professional indemnity insurance and clients' money should make clients feel reasonably confident about entrusting their money to chartered surveyors in their professional capacity. Unfortunately, however, rules are no guarantee against dishonesty, and the Institution, in common with other professional bodies, has an extra safety net to protect clients against loss caused by dishonesty where the money is not otherwise recoverable.

Under this scheme, the Institution itself undertakes to reimburse clients their direct pecuniary loss in certain circumstances and up to fixed limits. The details are summarised in Appendix VII.

The Institution itself insures this risk and recovers the cost of the insurance premium and associated administration from Member principals in private practice who do not submit 'nil' returns under the Members' Accounts Regulations (i.e. those firms who do hold clients' money). The clients' money protection scheme 'levy' may vary from year to year. At present it is £20 for each chartered surveyor liable to pay it.

To reduce administration costs, the Institution charges the levy automatically to Members listed on its records as partners, directors or sole principals of firms or companies providing surveying services *unless 'nil' returns have been received from those Members*. It is, therefore, up to Ian to ensure that, if his firm did not hold clients' money, a 'nil' return is made in time, otherwise he may be charged the levy.

One question that is often asked about the clients' money protection scheme is 'why must I pay the levy when my firm is already insured against losses arising from the dishonesty of partners and employees? Am I not paying twice?' It is true that the Institution's Rules require Members' professional indemnity insurance policies to include dishonesty cover. But there are circumstances where that cover would be invalid, for example, where a partner himself committed or connived at the dishonest act resulting in the loss. No-one can insure against his own dishonesty.

To the vast majority of chartered surveyors who are honest, I would only add that the levy is a small price to pay for the good name of the profession.

E Connected businesses: Bye-Law 24(2) and Regulation 6

'No Member shall be connected with any occupation or business in any way which would, in the opinion of the General Council, prejudice his professional status or the reputation of the Institution.'

In the past, previous versions of this Bye-Law sought to prevent chartered surveyors from engaging in commercial activities, which were thought to be incompatible with generally accepted notions of professionalism. Times change, however, and commercialism and competition are now essential elements of a successful practice – provided that professional ethics are preserved. Similarly, it is no longer considered strange for professional services to be combined with services which were not traditionally associated with the established professions, for example, a removal service, insurance or travel agency. In fact, more and more clients expect, or find it convenient, to look to one firm or company for several associated needs.

These changes call for vigilance from professional people to ensure that, when offering any of the multitude of 'commercial' services that they are now allowed to offer, they do not inadvertently harm their clients' interests, or the reputation of their profession. Bye-Law 24(2) remains, therefore, in a revised form, to enable the Institution to take action against chartered surveyors who step over the line of what is currently acceptable in combining occupations or services under the banner of surveying.

The Bye-Law is now normally invoked only with Regulation 6, which deals with specific situations:

> '6. *Without prejudice to the generality of Bye-Law 24(2) a Member carrying on practice as a Surveyor:*
>
> (a) *shall not be permitted to take part directly in the management or control of any organisation whose business consists wholly or substantially, or in the case of a Member performing the function of quantity surveying, wholly or partly, of building or civil engineering contracting;*
>
> (b) *shall be permitted to conduct, or allow to be conducted, as an adjunct to or in conjunction with his practice as a surveyor any business (a 'connected business') which in the opinion of the General Council is of such a nature that neither the Member's professional status nor the reputation of the Institution is thereby prejudiced;*
>
> (c) *in any case where the General Council is satisfied that a connected business is of such a nature or has been conducted in such a manner as to prejudice a Member's professional status or the reputation of the Institution, it may:*
>
> (i) *prohibit that Member from conducting or allowing to be conducted such business, or any similar business, as a connected business and/or*
>
> (ii) *deal with that Member in any one or more of the ways specified in Bye-Law 25(1).'*

Regulation 6(a) prohibits chartered surveyors who are in private practice from involving themselves in the building industry in ways which might disadvantage their clients or destroy the confidence that contractors and clients need to have in the independence of professionals.

The Institution has in the past indicated the factors which would be taken into account in determining the level of direct involvement in a contracting company. These include the size of a Member's holding and his position in the company. At Appendix VIII, I have reproduced extracts from guidance published by the Institution on the operation of Regulation 6.

Regulations 6(b) and (c) allow chartered surveyors to link their practices with non-surveying businesses so long as neither the Member nor the Institution is brought into disrepute. The sort of links which are envisaged here could include shared premises or staff; financial connections; shared directorships; a common house-style or corporate image. The sort of businesses with which chartered surveyors might wish, and would now be allowed, to link their practices include removals services, travel agencies, publishing and data processing, software and financial services.

Ian's practice would, therefore, be allowed to offer any of these services, or establish links with other companies doing so. However, if the Institution were to receive complaints that these non-surveying services were being conducted badly or improperly, it could (under Regulation 6(c)) either discipline Ian and his partners or prohibit them from further involvement with the services, or both.

F Conflicts of interest: Bye-Law 24(3) and Regulation 8

A conflict of interest may arise either between the interests of two or more clients of a surveyor, or between the personal interests of a surveyor and the interests of a client.

Bye-Law 24(3) says:

'Subject to the Regulations, it shall be the duty of every Member:

(a) when acting for a client or when in contemplation of acting for a prospective client whose interests conflict or may conflict with his own or those of any of his associates as defined in the Regulations, to disclose the relevant facts forthwith to the client or prospective client and to the associate and where such disclosure is oral to confirm the same in writing at the earliest opportunity and inform the client that he will be unable to act or continue to act unless the client requests him so to do after obtaining independent professional advice;

(b) *to ensure that neither he personally nor any firm or company carrying on practice as surveyors of which he is a partner or director acts for two or more parties with conflicting interests without disclosing the relevant facts to each of those parties forthwith and confirming such disclosure in writing at the earliest opportunity.'*

The relevant Regulation directly having a bearing on conflicts of interest is Regulation 8 although other Regulations, while not directly aimed at conflicts of interest, do touch upon it, for example, notifying client of terms and conditions (Regulation 2), stipulations that the surveyor is to act for the purchaser (Regulation 3), connected businesses (Regulation 6), Company employees offering surveying services to third parties (Regulation 7), taking financial interest (Regulation 9) and trade discounts (Regulation 22).

Regulation 8 provides as follows:

'(1) *Without prejudice to the generality of Bye-Law 24(3) where a Member is instructed:*

(a) *to sell or let a property and it is intended that the property be acquired by the Member himself or an associate of his, or*

(b) *to negotiate the purchase or lease of a property owned by the Member himself or an associate of his*
he must

(i) *make full and immediate disclosure to his client of all the relevant facts;*

(ii) *make a declaration of his interest to the vendor's or lessor's, or purchaser's or lessee's, Solicitors, as the case may be;*

(iii) *inform the client that he will be unable to act or continue to act unless the client requests him so to do after obtaining independent professional advice; and*

(iv) *confirm the requirements of (i), (ii) and (iii) above in writing at the earliest opportunity.*

(2) *Where a Member acts as agent for the sale or letting of property owned by the Member himself or associate of his, or in which he or an associate of his has an interest, he must disclose the relevant facts to the prospective purchaser or lessee or their solicitors.*

(3) *Where a Member conducts a connected business as permitted by Regulation 6(b), or where he allows such a business to be conducted, and where the interests of a client of the practice conflict or may conflict with the interests of a client or customer of the connected business, he must disclose the relevant facts forthwith to all parties concerned and confirm the disclosure in writing at the earliest opportunity.*

(4) *In Bye-Law 24(3) and this Regulation 'associate' may include the Member's firm or company or a partner of his in the case of a firm or co-director in the case of a company, his spouse, a near relative of his by blood, adoption or marriage, his nominee or employee, or any other person, firm or company so associated with the Member that his or its interest may conflict with that of the client.'*

As you will have already deduced, under the Institution's Rules there is no absolute bar on acting when there is a conflict of interest, there is only a duty to *disclose* that interest.

If you are faced with a situation where a conflict of interest has arisen or might reasonably be expected to arise, then you should immediately make disclosures. It is not sufficient merely to telephone the client and advise him of the conflict or potential conflict. The disclosure should be confirmed *in writing* at the earliest opportunity and disclosure should be of all the relevant facts. The reason for this is clear; if at a later stage allegations of impropriety are made then a Member can produce his copy of the written disclosure to refute those allegations.

Regulation 8(1)(b)(iii) requires the Member to notify the client that the Member will not be able to act or continue to act 'unless the client requests him to do so *after obtaining independent professional advice*'. The question arises as to how far the Member should go in making sure that his client has received 'independent professional advice'? Having made the disclosure and confirmed it in writing it would be prudent not to act or continue to act until in receipt of instructions from the client acknowledging the written disclosure, confirming that he has obtained independent professional advice, and requesting the Member to act or continue acting. A Member acting or continuing to act without such written confirmation from the client still risks the possibility that his client may allege impropriety and deny receipt of any written notification of the disclosure.

You may be led to assume that conflicts of interest arise only in the practice of estate agency. They don't; *conflicts can arise in all areas of practice.* I am sure that if you take a few moments to consider your own field of practice, you can envisage circumstances when a conflict of interest may arise.

It is without doubt true that the vast number of complaints made about conflicts of interest arise out of estate agency. I make no apology therefore for giving the following hypothetical examples. They are based on those contained in Volume 17 of *Chartered Surveyor Weekly*, RICS News at page 623.

Structural Surveys

Example 1

Let us assume that Carefree, Jolly & Happy together with Anytown Mega Estates Limited and another local agent, Sell Sell & Sell holds direct instructions from Mr. Smith, the owner of 99 The Street, Anytown, to sell that property. A prospective purchaser, Mr. Black, seeks to instruct Ian's practice, Carefree, Jolly & Happy, to carry out a full structural survey of the property.

Carefree, Jolly and Happy *may* accept Mr. Black's instructions provided that it makes full disclosure to both Mr. Black and to Mr. Smith so that they understand that Carefree, Jolly & Happy will be acting for both of them and provided also that Carefree, Jolly & Happy confirms those disclosures in writing as soon as possible. However, there are very good reasons why in such circumstances Members *should decline* Instructions. For example:

- the vendor client, Mr. Smith, would probably be very upset indeed to think that an agent he had instructed to act on his behalf, in the disposal of the property, was now going to act on behalf of the prospective purchaser;
- even if the vendor client was not unhappy at the thought of his agent undertaking a full structural survey on behalf of the purchaser, his attitude would be likely to change if the

purchaser subsequently withdrew as a result of an adverse survey.

Even though the Institution's Rules themselves do not prohibit a Member from acting in those circumstances, common sense dictates that it would be very unwise to do so. If a Member did decide to act for both parties and made the disclosure required by Bye-Law 24(3)(b) he would be well advised to obtain written confirmation from both parties that they were content for him to act in the circumstances. If he does not, then there could be further problems if, for example, Mr. Smith does not really address his mind to the issues when the disclosure is made to him verbally and then takes exception at a later date, when he receives the letter of confirmation. By this time events may have progressed. If, between speaking to Mr. Smith and Mr. Smith's receipt of the disclosure letter from Carefree, Jolly & Happy that it has carried out the survey, Mr. Smith might claim that this constituted a trespass.

Example 2

Let us now assume that Carefree, Jolly & Happy, having followed the steps above, re-accepts Mr. Smith's instructions to sell the property. Let us also assume that it finds a prospective purchaser, Mr. Green, who learns, from whatever source, that Carefree, Jolly & Happy has recently carried out a structural survey; Mr. Green asks Carefree, Jolly & Happy about it.

Whether or not the report revealed any serious defect, and whether or not Mr. Smith has agreed to have the defect remedied, Carefree, Jolly & Happy will risk incurring liability for professional negligence if it makes a statement about the condition of the property which proves to be wrong. In addition, Carefree, Jolly & Happy could involve its client, Mr. Smith, in liability for misrepresentation; Mr. Smith could then also have a claim against Carefree, Jolly & Happy for professional negligence or for breach of the contract of agency. The only sure way to avoid these risks is for Carefree, Jolly & Happy to decline to re-accept Mr. Smith's renewed instructions to sell.

Sub Agency

Example 3

Assume in this example that Carefree, Jolly & Happy and Sell Sell & Sell hold instructions, not direct from Mr. Smith, but as sub agents to Anytown Mega Estates Limited.

There is no difference in the principle where an agent holds instructions as a sub agent and is the agent who is instrumental in introducing the prospective purchaser, from which action he expects to derive a share of the commission. In this case the duty to disclose his dual capacity to both the vendor, Mr. Smith, and the purchaser, Mr. Black, is the same. Normally it would be a simple matter for a sub agent in these circumstances to terminate his instructions by informing the main agent. However, the sub agent should not re-accept instructions in the event of the purchaser's withdrawal.

Building Society Valuations

Example 4

Again, let me take the example set out in 1 above, but assume that Carefree, Jolly & Happy is on the panel of valuers of the Anytown Building Society to whom a purchaser has applied for a mortgage.

In this situation there is a clear conflict of interest and Carefree, Jolly & Happy must comply with Bye-Law 24(3)(b) if it is to act. If Carefree, Jolly & Happy has introduced the purchaser to the property then it *must* refuse the Building Society's instructions. Section 13(3) of The Building Societies Act 1986 prohibits the Building Society relying on a mortgage valuation compiled by a valuer who is also selling the property, the subject of the valuation. Similarly, if Carefree, Jolly & Happy has another prospective purchaser in view, it should decline the Building Society's instructions to ensure that, however indirectly, no conflict of interest could possibly arise in the near future. It should decline instructions as soon as possible to avoid any subsequent suggestion that a delay in doing so had in some way helped its purchaser.

If Carefree, Jolly & Happy did not introduce the prospective purchaser, but continues to hold the vendor's instructions to sell, it should inform the Building Society immediately that it declines the Society's instructions. It should also notify the vendor as soon as possible. An alternative course, however, is that Carefree, Jolly & Happy or the vendor might terminate the relationship created by the instructions to sell, thereby avoiding the conflict arising from acting for two masters. If this alternative was followed, Carefree, Jolly & Happy would have to decline any renewed instruction to sell by the vendor. In all these circumstances early contact is essential. Carefree, Jolly & Happy should seek to make contact by telephone in the first instance, although of course it must subsequently notify the vendor in writing. The requirements of Bye-Law 24(3)(b) are only satisfied when Carefree, Jolly & Happy has informed both the vendor and the Building Society of the relevant facts in writing.

There may be practical difficulties in contacting the vendor, either because the report is required at short notice, or because the vendor is not actually living in the property. In these circumstances, Carefree, Jolly & Happy must make every endeavour to contact the vendor client. If for any reason it cannot make contact with him it must decline instructions to act for the Building Society. However, if Carefree, Jolly & Happy reasonably believes that in the ordinary course of post a letter will reach the vendor by the time it intends to carry out the valuation, then it need not delay the work to wait for an acknowledgement from the vendor that he has received the letter *but it would be well advised to do so*.

In cases involving conflicts of interest in estate agency, the provisions of Section 21 of the Estate Agents Act 1979 will be relevant. As elsewhere in this guide I have assumed that the reader is conversant with that Act.

To sum up, therefore, if there is a conflict or a reasonable probability of a conflict arising, Members must comply with the disclosure provisions of the Institution's Rules and comply with any statutory provisions, some of which go further and *prohibit* Members from acting. The only safe answer may be to decline instructions.

G Confidentiality of clients' business: Regulation 19

When instructing his professional advisers a client must necessarily divulge information which he would not wish to be made public. He is entitled to expect that his adviser will respect his confidentiality and not pass on such information.

Regulation 19 states that:

'(1) No Member shall, without his client's consent, disclose personal information concerning that client.

(2) For the purposes of this Regulation 'personal information' shall include any information relating to the personal, financial or business circumstances of any person or body, or any other information of a sensitive nature which, having been disclosed to the Member in his professional capacity, is not known to have been otherwise generally published.'

A breach of this Regulation would almost inevitably amount also to a breach of Bye-Law 24(1).

H Claims of superiority: Regulation 18

Members should always think twice before claiming that their firms are better (or quicker or bigger) than other firms. Regulation 18 provides that:

'No Member shall make a public statement claiming an advantage or superiority over other firms, unless such a statement can be substantiated to the satisfaction of the General Council.'

This Regulation now allows claims of superiority, which were for many years prohibited outright. However, if a member of the public or another surveying firm challenges the claim, the Member making it will be obliged to withdraw it if he cannot substantiate it factually in disciplinary proceedings.

Carefree, Jolly & Happy cannot simply state in all their advertising literature 'We are the best'. However, if the partners believe themselves to be better than others in particular ways, they may make factual statements about their advantages, provided that they can substantiate the statements if challenged. For example:

(i) **'The only local Agents open 7 days a week'**
(The partners must ensure that the statement stops appearing the moment they learn that another local agent is also open for seven days.)

(ii) **'Offering the widest range of services in Anytown'**
(Do they? Still?)

(iii) **'Anytown's only independent Surveyors'**
(What does independent mean? Are the partners sure no-one else has opened an office locally?)

(iv) **'We aim to offer the highest standards'** (or 'widest choice', 'best advice' etc.)
(This is a fairly safe formula: a statement about what the firm *aims* to offer is not open to challenge in the way that a *claim* to offer the highest standards can be. It can also be a useful way of motivating staff to conduct themselves professionally.)

I Vicarious liability: Bye-Law 24(5)

I have already mentioned Ian's and Ann's obligations regarding other people's conduct when they were employees. As a partner, Ian can be disciplined by the Institution for the conduct of *anyone* in his firm, whether or not they are Members. The full text of the relevant Bye-Law, Bye-Law 24(5) reads:

'(5) (a) For the purposes of this paragraph (5) of this Bye-Law
(i) 'Partner' shall include a sole principal of a practice or a partner in a firm or a director of a company;
(ii) 'firm' shall mean a practice, firm or company carrying on practice as surveyors;
(iii) 'contravention' shall include any act or omission which would, if committed by a Member, constitute a contravention.

(b) Every Member who

(i) is or holds himself out or allows himself to be held out to be a Partner in a firm; or

(ii) allows his name and/or designatory letters or designation to appear on the notepaper or in the advertisements of any firm in which no partner is a Member; or

(iii) is a partner in a firm which is so connected with another firm, in which no partner is a Member, that in the opinion of the General Council persons dealing with that other firm believe or may be induced to believe that the two firms are connected;

shall be held responsible for any contravention of the Bye-Laws or the Regulations committed by any partner or by any member of the staff of such firm or other firm, provided that if such Member shall show that without any default on his part he had no reason to be aware and was not aware of any such contravention and that he had prior to such contravention taken all such steps as may be reasonable to ensure that such contravention was not committed he may be acquitted of responsibility for such contravention.'

Paragraph (b)(i) concerns partners (and Members who allow themselves to be 'held out' as partners – see Section 8.1A above) and makes them responsible for *any* contraventions of the Rules by *anyone* in the firm, unless they can show that they had 'no reason to be aware' and were 'not aware of any such contravention' and that they 'had prior to such contravention taken all such steps as may be reasonable to ensure that such contravention was not committed.'

Paragraph (b)(ii) places similar responsibilities on Member employees who are mentioned on the notepaper or in advertisements of firms which have no Member partners, as we have seen in Section 8.1 above.

Paragraph (b)(iii) extends Ian's responsibilities as a partner in his own firm to the conduct of partners or staff in another firm that appears to be connected with his. If Ian's firm should join a group of firms and benefit from the wider publicity that might result, *he* could be disciplined by the Institution for breaches of any Rules committed by non-Members in any of the other firms which have no Member principals.

As far as non-Members in associated firms are concerned, Ian should ensure that those people in authority are familiar with and understand the Institution's Rules and that they are prepared to comply with them.

What should Ian do in any of these situations, to take 'reasonable' steps to prevent contraventions by others, so that he would have a defence before the Institution's Professional Conduct Committee if he were unfortunate enough to be taken to task for the misconduct of others?

First, in relation to his own employees, he should ensure that all staff have a provision in their contracts of employment:

(1) acknowledging receipt of a copy of the Institution's Bye-Laws and Regulations;
(2) requiring staff to familiarise themselves with them; and
(3) requiring staff to seek his advice or that of another Member partner if they are in doubt about the application or interpretation of the Rules.

Next, Ian should ensure that there is a clear chain of responsibilities within the firms for extracting from the Journals of the Institution explanatory articles on the Rules and notices of amendments and for circulating these throughout the practice.

Also, Ian and his partners should ensure that one partner within the practice assumes responsibility for dealing with complaints received by the practice, whether from clients or from the Institution. If the nominated partner is also made responsible for professional indemnity insurance matters he will know precisely what matters need to be referred to his insurers. Often complaints from clients involve suggestions of negligence and it is sensible that at least one partner should be aware of all matters involving allegations of breaches of the Rules or possible insurance claims. If one partner is nominated to investigate and deal with complaints it will help to ensure that they are dealt with consistently and promptly.

These precautions would constitute 'reasonable steps' in most circumstances. But whether the Institution would regard them as *sufficient* to acquit a Member of responsibility for the misconduct of others will depend on all the circumstances of a case. Members should take whatever additional steps they can to minimise the risk of misconduct by their colleagues, staff or associates.

9.3 Partnership name

Carefree, Jolly & Happy has been trading for some time. However, until Ian Hopeful was admitted into partnership the name of the firm consisted of the surnames of all the partners. Now this is no longer the case and the partnership is caught by Section 1 of the Business Names Act 1985. The full text of Sections 1–7 is set out at Appendix XI.

The Business Names Act 1985 requires that where anyone carries on business under a name other than his own, he must state on all his business literature the full name of the business and the place within Great Britain where any document relating in any way to the business may be served. The practice of Carefree, Jolly & Happy was not required to do this when Messrs. Carefree, Jolly and Happy were the only partners. Now that Ian has joined them, and since his name is not incorporated in the title of the business, it is necessary for all the partners' names to be stated in legible characters on all business letters, written orders for goods or services to be supplied to the business, invoices and receipts issued in the course of the business and written demands for payment of debts arising in the course of business. The Act also requires (in Section 4) that on any premises where the business is carried out there must be displayed, in a prominent position so that it can be easily read by customers or suppliers, a notice containing the names of the partners and an address for service of documents.

Carefree, Jolly & Happy will, therefore, have to produce such a list and make sure that it is prominently displayed in its offices. It will also have to check its literature to make sure that Ian's name appears on it where the names of the other partners appear. Partnerships which fail to comply with the requirements of the

Business Names Act commit a criminal offence under Section 7. They may also find themselves in difficulty if they seek to bring legal proceedings while they are in breach of the Act.

9.4 Partnerships and the chartered designation

Ian's practice is entitled to use the designation 'Chartered Surveyors' in connection with its title because each individual partner of the firm is either a Fellow or a Professional Associate. However, Ian and his two partners, Carefree and Happy, are still Professional Associates and they must, therefore, continue to show the initials appropriate to their class of membership (ARICS).

Bye-Law 5, which is reproduced at Appendix II governs the use of the chartered designation. Members who have been elected Fellows are entitled to use the initials, 'FRICS'. Professional Associates are entitled to use the initials, 'ARICS'. All such Members may call themselves a 'Chartered Surveyor', but Bye-Law 5(3) provides for alternative designations.

Ian and his partners are Members of the General Practice Division of the Institution. They may describe themselves as 'Chartered Surveyors' or they may choose to use the alternative designation, 'Chartered Valuation Surveyors'. They are also permitted to add to that designation the words 'and Estate Agents'.

A study of the history of the Institution will reveal that it has, over the years, merged with numerous other professional bodies, including the Chartered Land Agents Society and the Chartered Auctioneers and Estate Agents Institute. Older Members who were entitled to use the designatory letters of those bodies are now only entitled to use the designatory letters 'FRICS' and 'ARICS'. Use of the designatory letters 'FLAS', 'QALAS', 'FAI', 'AAI', 'FIQS' or 'AIQS' are no longer permitted.

There are two further categories of Members: Associates and Honorary Members. The Institution has the power to elect up to 50 Associates who are entitled to vote and take part in the affairs of the Institution but who are not, by profession, surveyors. Their pursuits must be such as to qualify them to concur with surveyors in the advancement of professional knowledge. Current Associates include distinguished members of the legal profession and of the judiciary. There is provision for one Associate Member to sit on the General Council of the Institution. 'Associates' must not be confused with 'Professional Associates'. The latter are corporate Members who are entitled to use the designatory letters 'ARICS'.

The other category of membership of the Institution is Honorary Membership. Honorary Membership is conferred upon persons who, by reason of their position, experience or eminence may be able to render assistance in promoting the objects of the Institution but who do not practice as surveyors in the United Kingdom of Great Britain and Northern Ireland or in the Republic of Ireland. They are not entitled to be present or take part in any Extraordinary General Meeting, but do have the privilege of being present at and taking part in all Annual and Ordinary General Meetings. They do not have the right to vote on any Resolution proposed at an Annual General Meeting. The Institution is currently honoured to include amongst its Honorary Members His Royal Highness the Duke of Edinburgh, His Royal Highness the Prince of Wales and His Royal Highness the Duke of Gloucester.

Associates and Honorary Members are not entitled to use any distinguishing letters after their name. However, it is proposed to amend Bye-Law 5 to entitle Honorary Members to use the distinguishing letters 'Hon. Memb. RICS' and entitle Associates to use the distinguishing letters 'Hon. Assoc. RICS'. These proposals are to be voted upon at the next Extraordinary General Meeting in October 1990.

I have explained above (in Section 8.1) how Ian and Ann must be very careful to ensure that their status within their practices was clearly stated while they were employees. I also quoted Regulation 13 in full.

You will recall that it is incumbent upon Ian, now that he is a partner, to make sure that if any chartered surveyor employees are named on the firm's literature, that their status within the firm is clearly stated.

10. Directors

Let us now see how Ann Bright is getting on. She has agreed with the directors of Anytown Mega Estates Limited to accept a seat on the Board. The notepaper has been re-printed as in Figure 7. What are her responsibilities now?

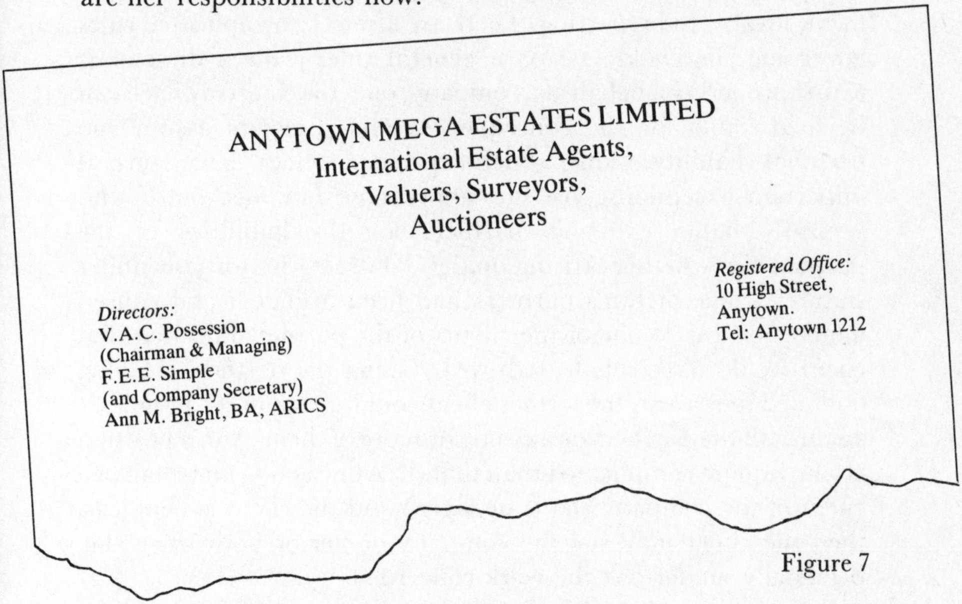

ANYTOWN MEGA ESTATES LIMITED
International Estate Agents,
Valuers, Surveyors,
Auctioneers

Registered Office:
10 High Street,
Anytown.
Tel: Anytown 1212

Directors:
V.A.C. Possession
(Chairman & Managing)
F.E.E. Simple
(and Company Secretary)
Ann M. Bright, BA, ARICS

Figure 7

10.1 Directors' legal responsibilities

Directors are the agents of the shareholders of the company appointed to look after the business on the shareholders' behalf. In one famous case, Lindley LJ said:

> 'Directors are not only agents, but to a certain extent trustees for the company and its shareholders . . . they are not the masters but the servants of the shareholders; the power of a director is limited and accompanied by a trust, and is to be exercised bona fide for the purpose for which it was given, and in the manner contemplated by those who gave it . . . so that the powers which the directors have . . . are reposed in them in order that the

> powers may be bona fide exercised for the benefit of the company as a whole; and any exercise of such powers for any other purpose is regarded as a breach of trust, and is treated accordingly.'

Their rights, duties and liabilities are largely enacted in the Companies Act 1985 and include duties of disclosure, management, loyalty and reporting. There are also very complicated rules governing shareholdings. As a general rule, while a director is acting properly on behalf of a company, only the company itself can be held liable for the consequences and directors assume no personal liability. This general principle does have several important exceptions. We saw above how Ian was jointly and severally liable with his partners for the liabilities of the partnership whether arising under contract, in tort or under statute. If one of Ian's partners had been negligent and caused damage or loss to one of the clients of the partnership, then that client would have sought redress by suing the partners. Having obtained judgment, the former client could then enforce judgment against all the partners or any one or more of them. Ann's position is a little more complicated than that. If, while acting on behalf of a client of the company she is negligent and the client suffers loss, then the client may sue the company or her or both since she, personally, undertook the work concerned.

What if the negligence which caused the loss to the former client was not Ann's, but that of an employee working under the supervision and direction of Ann? It might, for example, have been a student working for the company as Ann's assistant who was negligent. In that case, it is highly probable that the former client would have a right of action against, not only the company and the employee responsible for the negligence which resulted in the loss or damage, *but also against Ann herself*, being the director of the company directly supervising the work of the employee.

Directors can also be personally disqualified from holding office. The Companies' Act 1985, the Company Directors' Disqualification Act 1986 and the Insolvency Act 1986 have greatly increased the powers of the Court to disqualify individuals from acting as directors. The period of disqualification can be up to 15 years and, during that period, a person so disqualified cannot hold office as a director, promoter, liquidator or administrator of a company. A former director acting in breach of a Disqualification Order not only faces fines of up to £2,000 and imprisonment up to six months, but is also personally liable for the debts of the company he manages while he is disqualified.

A director can be disqualified for:

(i) having been convicted of an indictable offence concerning the promotion, formation, management or liquidation of a company or with the receivership or management of a company's property (Section 2 Company Directors' Disqualification Act 1986);

(ii) persistent default in relation to the provisions of the companies legislation requiring any return, account or other document to be filed with, delivered or sent or notice of any matter to be given, to the Registrar of Companies (Section 3 Companies Disqualification Act 1986);

(iii) fraud or fraudulent trading;

(iv) wrongful trading;

(v) acting as a director or taking part in or being concerned in the promotion, formation or management of a company without leave of the court by an undischarged bankrupt.

The Insolvency Act itself also introduces new offences:

- destroying or falsifying companies' books;
- fraud in anticipation for winding up of a company;
- making a material omission from a statement relating to the company's affairs.

10.2 Directors' professional responsibilities

Now that Ann is a director she will be required to comply not only with the Rules that were relevant to her as an individual (Section 1) and as an employee (Section 2.2) but also to:

(1) effect Professional Indemnity Insurance to comply with Bye-Law 24(9A) and the Compulsory Professional Indemnity Insurance Regulations;
(2) make returns under the Members' Accounts Regulations;
(3) pay the Clients' Money Protection Scheme levy unless she makes a 'nil' return under the Members' Accounts Regulations;
(4) comply with Bye-Law 24(2) and Regulation 6;
(5) take direct responsibility for ensuring the company handles conflicts of interest properly (Bye-Law 24(3)) and Regulation 8;
(6) take direct responsibility for the company's publicity;
(7) take direct responsibility for the handling of confidential information disclosed to the company by its clients; and
(8) ensure that her co-directors and members of staff are familiar with and do not contravene the Institution's professional rules. (Bye-Law 24(5)).

These responsibilities are described in full in respect of Ian as a partner (see Section 9.2). As a director of a company Ann has an extra duty. She must ensure that the Memorandum and Articles of Association of Anytown Mega Estates Limited include the clauses required by Regulation 10.

Regulation 10(1) requires that:

'A Member may carry on practice as a surveyor through the medium of a company provided that he complies with these Regulations.'

Regulation 10(2) provides:

'No Member shall carry on practice as a surveyor as aforesaid in any case where he is a director of the company concerned unless:

(a) a provision to the following effect is included in the Memorandum of Association or equivalent constitutional document of that company and in such manner that it qualifies those powers of that company enabling it to offer surveying services:

Any business of surveying for the time being carried on by the company shall at all times be conducted in accordance with the Rules of Conduct for the time being of The Royal Institution of Chartered Surveyors.';

(b) *a provision to the following effect is included in the Articles of Association or equivalent constitutional document of that company:*

It shall be the duty of the Directors to ensure that any business of surveying for the time being carried on by the Company shall at all times be conducted in accordance with the Rules of Conduct for the time being of The Royal Institution of Chartered Surveyors.'

It is worth pointing out that Ann will be responsible for the conduct of the company whether or not the Regulation 10 clauses appear in the Memorandum and Articles. Failure to include them will, however, in itself put her at risk of disciplinary action by the Institution. The Regulation is there largely to support Ann in her endeavours to ensure that her co-directors, as well as she, enforce the Institution's Rules throughout the company. If they do not, they will be in breach of their company's Memorandum and Articles and will also be in trouble with their own shareholders.

In addition, under Bye-Law 24(5) Ann becomes vicariously responsible to the Institution for the conduct of anyone in the company, whether a chartered surveyor or not, as Ian is as a partner, and as either of them would have been if they had allowed themselves to be called 'Associate Partner' or 'Area Director' before their promotion to principals.

Ann is probably in a more difficult position than Ian. At least Ian's partners are chartered surveyors and ought to know what the Institution requires of them so far as their conduct and the conduct of their practice is concerned. Ann's co-directors are not chartered surveyors and her first action must be to make sure that they are familiar with the Institution's Bye-Laws and Regulations. She must also ensure that all staff employed by the company are familiar with the Rules and that they strictly adhere to them.

There is one former restriction with which Ann no longer needs to concern herself. Since October 1986 the Institution's Rules have placed no limit on the amount of a surveying company's shareholding that may be held by persons or bodies other than the directors. It is, therefore, permissible for her to be a director of a company which is wholly owned by another or for its shares to be floated on the Stock Market, in the case of a public limited company.

10.3 Company names

Except where private limited companies are exempt from the requirement (see below), all companies must add to their registered names the words 'public limited company' or 'plc' or 'limited' depending on whether or not they are registered as public companies. There are two exceptions, namely:

- a private company limited by guarantee; and
- a company which on 25 February 1982 was a private company limited by shares which held a licence under the Companies Act 1948 permitting the use of its name without the addition of the word 'limited'.

Companies need not trade under their registered or corporate name; they may use a trading name in which case the Business Names Act 1985 will apply to it. (See Appendix XI).

10.4 Companies and the chartered designation

In the case of companies it is necessary for all the directors to be chartered surveyors before the chartered designation may be used in conjunction with its corporate name.

Anytown Mega Estates Limited are not entitled to use the designation 'chartered surveyors' because, even though Ann is a chartered surveyor and may be a director, the other directors are not chartered surveyors. (See Appendix II for the full text of Bye-Law 5.)

Let us, however, look at the position if all Ann's co-directors were also chartered surveyors. Since January 1990 Bye-Law 5 allows a surveying company to use the chartered designation in conjunction with its corporate name if all the directors are chartered surveyors – even if some or all of the shares are held by people other than the directors. These 'outside investments', which have not been restricted by the Institution's Rules since October 1986, no longer even prevent the company from using the chartered designation if it is otherwise entitled to it.

I should stress that the chartered designation may only be used with the company's *corporate name*, i.e. the registered name. It may *not* be used with a *trading name* if it is different from its corporate name. This means that where a company's trading name appears prominently at the top of the letterhead, and the corporate name discreetly at the foot, it is only at the foot where the chartered designation may appear (see Figure 8).

One last point I must make about the use of the chartered designation by companies is that the words 'Chartered Surveyors' may *not* be registered as part of the company's corporate name. The Bye-Law permits their use (where all the directors are Members) *in conjunction with*, but not as part of, a company's name.

11. Sole Principals

Let us now assume that Ian Hopeful decides to separate from his partners and set up a practice on his own. He may have quarrelled with them; personal circumstances may have necessitated his moving away from the area; or he may simply feel that he has gained enough experience in a responsible position to 'do things his way'.

11.1 General responsibilities

A surveyor practising on his own account as a 'sole trader' is in a very responsible position. He must comply with the Rules already explained. In particular he is required to obtain Professional Indemnity Insurance under Bye-Law 24(9A), to comply with the Members' Accounts Regulations and to continue paying the Clients' Money Protection Scheme levy unless he makes a 'nil' return (see Section 9.2D). Under Bye-Law 24(5) he is responsible for the actions of his staff which might breach any of the Institution's Bye-Laws or Regulations. If he carries on practice

MEGAMOVE
100 High Street
Anytown
International Real Estate Agents – Valuers – Surveyors
Telephone Anytown 7000 Fax Anytown 2000 Telex 3000
Part of the Megamove Group with offices throughout the world

Megamove is a trading name of Anytown Mega Estates Limited, Chartered Surveyors,
Registered Office: 10 High Street, Anytown, Registered in England Number 123456789
Directors: A.M. Bright, ARICS I.M. Hopeful, ARICS

Figure 8

under any name other than his own or with any additions to his own name, he also needs to concern himself with the provisions of the Business Names Act 1985. If he chooses to incorporate his practice, he must of course include the Regulation 10 provisions in the Company's Memorandum and Articles.

11.2 Professional Indemnity Insurance

Professional indemnity insurance can be a very heavy financial burden for a sole practitioner. It is nevertheless essential – not only as a requirement of the Institution, but also for his own protection. Professional indemnity insurance is unlike any other type of insurance. Motor insurance, buildings insurance, contents insurance, health insurance and life insurance all insure against the possibility of some *future* event. Professional indemnity insurance is written on a 'claims made' basis and covers the *past* activities of the insured. It might be easier to understand if I give some examples.

Suppose that John started practice on his own account as a surveyor in 1970. He is very professional, conscientious and diligent in his approach to work and by 1985 decides that he is no longer going to effect professional indemnity insurance because the premiums are too high. He has never had a claim for negligence and is anxious to cut down his annual expenditure. In 1987 a former client of his writes to him pointing out that substantial cracks have appeared in the load-bearing walls of his house which John surveyed for him in 1983. John has no current professional indemnity insurance. He is sued by his former client and has to pay the damages and expenses himself. He cannot now buy professional indemnity insurance and expect his insurers to pay for a claim which has already been made. No insurer would cover him for claims already made. His previous insurers will not indemnify him because the claim has arisen *after* the period of insurance has expired even though he was insured at the time he is alleged to have been negligent.

Let me take the reverse of that situation.

Peter starts work in 1970 and has not bothered to obtain professional indemnity insurance. In 1975, having discussed matters with his solicitor and accountant, he decides that it will be prudent to effect professional indemnity insurance and does so with effect from the 1st January 1976. Along comes a former client in 1977 and accuses him of having been negligent in 1972, when he was not insured. Because he took out cover in 1976 *and* renewed his cover for professional indemnity insurance on 1st January 1977 his insurers meet the claim and he pays nothing other than any excess he may have under the policy. He is covered because he has insurance *at the time the claim was made* even though he had no insurance in 1972 when he is alleged to have been negligent.

A third example shows why it is necessary to have insurance during each year of practice, as well as after practice ceases.

Susan leaves her job with a local authority in January 1970 in order to set up as a sole principal, so that she is free to raise a young family. She wisely takes out insurance cover from the first day she opens her practice and is unlucky enough to have a claim against her in June 1970 for work she did in March 1970. Her insurers advise her to defend the claim in court and she wins: there are no damages to pay to the plaintiff but the Judge awards her only half her costs. Happily, these are met by her insurers, except for the amount falling within her uninsured excess.

After 15 years (1985), Susan decides to return to salaried employment and to close her practice. To avoid finding herself in John's situation, she, again wisely, maintains her policy on a 'run off' basis. In 1987 a former client makes a claim against her for failing to notice woodworm in the roof timbers of his house which she surveyed prior to his buying it. Her insurers cover her because she has purchased 'run off' cover. Because she had very few claims against her while in practice, and those she had were unsuccessful, her insurers have charged her a modest premium. Indeed, had she had no claims at all, she would probably have been eligible for the Institution's small business scheme under which she can purchase insurance cover at very competitive rates.

As you read in Section 9.2A professional indemnity insurance has, of course, been compulsory for all chartered surveyors who have been sole principals, partners or directors of or consultants to practices, firms or companies carrying on practice as surveyors since 1st January 1986, even if they have since ceased practice, either to retire or to become employees.

What many people seem to forget is that on retirement, though ceasing to practice as surveyors, the risk that they may be sued for some past negligent act still continues. The period of liability for some past negligent act or omission can continue for many years. It does not die with the death of the practitioner. Claims can be made against his personal representatives, if, either he was a sole principal, or his former partners/directors split up after he leaves or for some other reason fail to maintain cover for his or their past work. For this reason indefinite 'run off' cover is *essential* as well as being *compulsory* for Members who retired on or after the 1st January 1986.

12. Consultants

It is common for a chartered surveyor partner to be invited to become a consultant to the practice on retirement. If Ian, having reached that stage, is described on the notepaper as 'Consultant Partner' then he is in the same position as if, previously, he had been described as 'Associate Partner'. He is liable in just the same way as when he was a partner, in contract, tort and under statute. If he is simply described as 'Consultant', then the law treats him as being an employee.

In some respects the Institution does so too. But in others it imposes obligations on him akin to those of a partner, director or sole principal.

12.1 Vicarious liability

If he is a consultant to a practice where no partner or director is a Member, Bye-Law 24(5)(b)(ii) makes him responsible for a contravention of the Bye-Laws or Regulations *committed by anyone in the practice*. (See Section 8.1C.) He can also be held responsible even if his name does not appear on the notepaper, but if his name or chartered surveyor status is mentioned in the firm's advertisements. A chartered surveyor consultant to a non-Member practice should therefore, ideally, obtain written confirmation from the principals either that he will never be mentioned on their notepaper or in their advertisements; or, better still, that they will endeavour to ensure that the Institution's Rules are observed throughout the firm.

12.2 Professional Indemnity Insurance

The Institution holds chartered surveyor consultants responsible for their own compliance with its Compulsory Professional Indemnity Insurance requirements. (See Compulsory Professional Indemnity Insurance Regulation 1 which includes amongst those Members to whom the Regulations apply 'a Consultant to a practice, firm or company carrying on practice as surveyors').

Where a chartered surveyor becomes a consultant to his former practice or company and the partners or directors include chartered surveyors, who maintain a professional indemnity insurance policy which complies with those requirements, the consultant will in most cases be covered under the policy. The consultant must, however, satisfy himself that this is the case since, if it is not, he as well as the Member principals could be disciplined by the Institution. He would also, of course, risk finding that he had no insurance to meet claims against him personally should he be unlucky enough to attract any.

A chartered surveyor consultant to a practice which has no chartered surveyor partners or directors should be even more careful to check his position, since the firm itself would not be required by the Institution to hold professional indemnity insurance cover. Even if the firm does have cover, it may not be as comprehensive as is required by the Institution, and he may not necessarily be covered under the firm's policy. He would then need to protect himself by obtaining his own cover.

It is not uncommon for a surveyor to become 'Consultant' to organisations and bodies other than his former practice. For example, a former client company for whom the surveyor has acted for many years may, on his retirement, appoint him as a consultant.

He may take two or three consultancies to supplement his pension. If he does so, he will be treated by the Institution as if he were a sole principal, maintaining his own small 'practice', and he would be required to effect a professional indemnity insurance policy to cover the risks to which he might be exposed by acting, or holding himself out, as a consultant to those organisations, even if his position was purely honorary and he did not give the company any professional advice.

Of course, a surveyor may accept a consultancy during his working life: as a partner in one firm, he may be offered a consultancy in another; or a regular client – perhaps a property company – may want his name on its notepaper as a consultant. Any chartered surveyor accepting such responsibilities must check with his insurers (and if necessary with the Standards and Practice Department at the Institution) that his current insurance covers the possibilities of claims arising out of such consultancy appointments. (There are occasional circumstances in which insurers have shown themselves reluctant to provide cover, for example, when a public authority or major company retains as a 'Consultant' a surveyor who was formerly an employee. The Institution's General Council may use its discretionary power to waive the Professional Indemnity Insurance requirements in such cases, but it would normally need to be satisfied that the Member had obtained appropriate indemnities from his employer/client. Any Member who has tried to obtain cover in these circumstances and failed should immediately seek advice from the Institution.)

12.3 Clients' money

Normally, the Members' Accounts Regulations will not apply to a consultant. However, if a chartered surveyor acts as a consultant and during his consultancy receives clients' money (as defined by the Regulations) in his own right as a sole principal, then he will be required to comply with the Regulations.

Chapter Three

CHARTERED SURVEYORS PRACTISING IN INDUSTRY, COMMERCE AND THE PUBLIC SECTOR

Now let us look briefly at the position of surveyors who pursue occupations outside private practice in industry, commerce and the public sector or as directors of property companies not offering surveying services to the public.

13. Holding out

The law relating to 'holding out' as I have briefly outlined earlier, will apply whenever a surveyor is employed. A surveyor who is employed, for example, by a development company should not allow himself to be described as 'Sales Director', 'New Homes Director' or 'Local Director' unless he is actually a director of the company in case he inadvertently binds the company beyond his authority.

14. The Institution's requirements

As far as the Institution is concerned, all chartered surveyors whether they are employed in private practice or outside it must comply with Bye-Law 24(1) and the other Rules listed in Chapter 1 Sections 4–7. I would also refer you to Section 8.1E – headed 'When is an employee not an employee?', since Members working outside private practice are often called upon to use their training and experience to advise others in their spare time.

14.1 Acting outside the course of employment

Let me remind you of the main provisions that will apply to an employed surveyor when acting as a surveyor outside the course of his employment:

A Clients' money

Whenever an employed surveyor acts for a client outside the course of his employment and receives money on behalf of a client, he is required to comply with the Members' Accounts Regulations because in that private capacity he is acting as a sole principal.

He should inform the Institution that he is doing so, and submit annual returns under Regulation 11(1)(a). If he does not submit a 'nil' return (to the effect that he has not held clients' money) he may be charged the Clients' Money Protection Scheme levy.

B Professional Indemnity Insurance

Similarly, when an employed surveyor acts in his private capacity for a client (whether for payment or not), he is required to be insured for breach of professional duty as a surveyor because, again, he is acting as a sole principal.

C Connected businesses: Regulation 6

Members in senior positions in the contracting industry must not engage in private practice at the same time (see Section 9.2E and Appendix VIII). Members in senior positions in other businesses who wish to engage in private practice 'in conjunction with' such other businesses need to be aware of the provisions of Regulation 6(b) and (c), and Regulation 8(3) (see Section 9.2E and F).

14.2 Acting within the course of employment

There are also a few provisions in the Institution's Rules which are relevant specifically to Members working outside private practice, in their capacity as such. While up to now I have consistently distinguished between employees and principals, including directors among the latter, the word 'employed' in the following two Regulations includes directors or chief executives of the non-surveying organisation envisaged. This is because the Institution's Rules do not distinguish between principals and employees when they are not 'in practice as surveyors'.

A Site boards: Regulation 5(3)(b)

'The Institution's standard nameboard may be used only;

(b) to exhibit the name and position of a Member employed by a public authority in respect of works carried out by that authority under his direct supervision, and where he is the senior officer concerned with such works.'

B Disclosure of employer's business: Regulation 7

'Every Member who is employed in a company or undertaking not primarily engaged in the provision of surveying services, and who in the course of such employment offers or provides surveying services to third parties, shall at the earliest opportunity make clear to the recipient or prospective recipient of such services:

(a) the identity of the Member's employer; and

(b) the nature of any business in which the Member's employer engages.'

Regulation 5 on Siteboards needs no amplification. Regulation 7 is aimed at ensuring that no recipient of professional services from a chartered surveyor is led to believe that the chartered surveyor is in a position to give him independent advice when in fact he owes his primary legal and professional duty to his employers. The Regulations made under the Financial Services Act 1986 similarly aim to ensure that recipients of financial services are quite clear whether they are being advised by independent or 'tied' advisers.

15. Legal liability

Perhaps one of the most important matters of concern for surveyors employed in industry, commerce and the public sector is the extent to which they can become professionally liable for work undertaken during the course of their employment. In the spring of 1990 an Institution Working Party reported on Professional Liability in the Public Sector and published a Memorandum of Guidance for Chartered Surveyors. Copies may be obtained from the Standards and Practice Department of the Institution. The working party identified the following main areas of risk for Members employed in the public sector:

- liability to third parties;
- liability to the employer;
- liability to another body for whom the employer is acting as agent;
- liability to an insurer (under the Insurer's rights of subrogation); and
- liability in respect of surcharge by external auditors.

The first four of those risks apply as much to surveyors employed in industry and commerce as to those in the public sector. A second Working Party has reported on Professional Liability in the Corporate Sector, and at the time of writing this has just published its findings. Copies are available from the Standards and Practice Department.

15.1 Liability to third parties

Where, during the course of his employment, an employee fails to take reasonable care to avoid acts or omissions which cause damage to others, he personally becomes responsible to those third parties for the consequences of his actions. His employer is vicariously liable and so an injured party will have a cause of action against either or both of them. Normally employers will hold professional indemnity insurance cover which will automatically cover employees while acting in the course of their employment. However, an employee should not naturally assume that he is covered by his employer's insurance (if the employer has such insurance) and a prudent employee will check with his employer to make sure that he is covered. If he is not, then he should seek legal advice on obtaining an indemnity from the employer.

15.2 Liability to the employer

Employees owe a duty to their employer to take care (*inter alia*) so as not to cause loss or damage to third parties. There is no implied term in a Contract of Service that an employee will be indemnified by his employer against such claims which expose him to risk. Therefore, if an employee in the course of his employment so acts as to cause loss or damage to a third party who subsequently seeks to recover that loss or damage against the employer, the employee will be liable to make good the loss or damage to the employer. This danger can be overcome by the inclusion in a Contract of Employment of a specific indemnity in favour of the employee.

15.3 Liability to another body for whom the employer is acting as agent

The liability of employees goes beyond liability to clients. For example, surveyors in the public sector are occasionally asked to undertake work on behalf of their employers' pension fund. The employee would be liable for his acts or omissions to that third party in just the same way as if the third party had been a client.

15.4 Liability to an insurer

After an employer's insurers have settled an insurance claim on behalf of the employer, the insurers are entitled to seek to recover their losses and expenses from the employee responsible for the loss or damage which resulted in the insurance claim under their rights of subrogation. Some professional indemnity insurance policies exclude rights of subrogation against employees, but *not all of them do*. Employees should ask to check the terms of their employer's indemnity insurance wherever possible and if there is a possibility that the employer's insurers will have rights of subrogation, the employee should seek an indemnity from the employer. In any case of doubt seek legal advice.

15.5 Countering the risks

The working party on Professional Liability in the Public Sector concluded that:

- *'Members must become more aware of the potential risks and take some action to ensure maximum protection against claims.'*
- *'It is in the best interests of both employers and employees that steps be taken to establish certainty and to remove concern.'*
- *'Members . . . must seek to obtain satisfactory arrangements through their normal means of negotiating terms and conditions. The RICS recommended form of Indemnity clause . . . gives basic protection.'*
- *Members '. . . are advised to check [their] employers' policy and investigate the possibility of reducing [their] own risk exposure by direct insurance'* and
- *Members should pursue these aims 'through their staff association and similar bodies.'*

While the working party's report concerned itself with Professional Liability in the Public Sector, the risks for Members which are identified are very similar to the risks to which Members employed in industry and commerce are exposed. They are very real and potentially very expensive. *No employed surveyor can afford to ignore them.*

Chapter Four

MISCELLANEOUS PROVISIONS

So far I have drawn attention to the main provisions in the Institution's Rules as they concern Members at different stages of their careers or in different capacities. I have not set out to produce a commentary on all the Rules, and in any event, there can be no substitute for reading the Rules themselves. They are given in full in the Appendices. However, before leaving the subject of professional Rules I would like to draw the reader's attention to the issues dealt with by the remaining Rules.

16. Offering gifts or inducements for the introduction of instructions

The offering of gifts or favours either to prospective clients or to third parties designed to secure instructions for work is prohibited by Bye-Law 24(6)(a), except in certain clearly defined circumstances. Regulation 14 *does* permit the payment of a fee or commission, or the provision of a gift or favour to a third party in recognition of the introduction of a client, if:

(i) *full disclosure* is made in writing to the prospective client *before* accepting instructions (except where the introduction has been effected in the ordinary course and/or by reason of the third party's surveying business); and

(ii) the Member has no reason to believe, and does not believe, that undue pressure or influence was exerted on the prospective client by the third party.

The words in brackets in sub-paragraph (i) mean that it would, for example, be permissible for Ann's practice to pay an introduction fee to Ian's practice for introducing a client to Anytown Mega Estates Limited without making disclosure in writing to the prospective client, because Ian's is a surveying practice. But if Ann's practice were to pay a commission to, say, the local postman for recommending her company then she would be required to disclose, in writing, the amount of the commission and the identity of the postman.

17. Quoting fees

Bye-Law 24(6)(b) prohibits Members from quoting a fee in competition for professional services without having sufficient information to be able to assess the nature and scope of the services required. But if a fee has been quoted by a Member, having obtained sufficient information to be able to assess the nature and scope of the services required by the client, the Member may not subsequently revise that quotation to take account of the fee quoted by another member of the surveying profession for the same services. (Bye-Law 24(6)(c).)

Neither may a Member quote a fee by reference to a fee quoted or charged by another member of the surveying profession. (Bye-Law 24(6)(d).) So to say to a prospective client, 'I will charge 10% less than any other local agent', would be a breach of the Rule.

18. Inviting instructions

Before the introduction of Regulation 20 the solicitation of instructions for work was not allowed except in very limited circumstances. (Bye-Law 24(7).) Chartered surveyors are now allowed to invite instructions for work provided:

(i) they do not do it by personal call to a private address; or

(ii) by telephone to a private address except in response to an advertisement of a property for sale or to let.

In the case of estate agency services, if a Member makes a personal approach for instructions he *must* give a clear statement to the prospective client:

(i) that if the prospective client has already appointed a sole agent, a liability to pay two commissions may arise if the second agent is appointed otherwise than as a sub-agent to that sole agent;

(ii) of the circumstances in which commission is required.

Where such statements are oral, they *must be confirmed in writing at the earliest opportunity*.

In the case of a Member inviting instructions for work other than estate agency, he must make a clear statement to the prospective client:

(i) that he will not accept instructions for work currently in hand and for which another professional adviser has been retained until the Member has first satisfied himself that the previous instructions have been terminated; and

(ii) the prospective client could, if a professional adviser were appointed to succeed the adviser already retained, be liable to pay fees to both professional advisers.

Any such statements *must be confirmed in writing at the earliest opportunity* (see Regulation 20 Appendix III).

19. Site boards

Regulation 5 deals with Members' site boards for use on a building site except where the Member is acting as an estate agent. Specifications as to the size of the board and size of the lettering are promulgated from time to time by the Institution and requests for the latest specifications should be addressed to either the

Professional Conduct Department or the Corporate Communications Department at the Institution.

20. Asset valuations

Chartered surveyors frequently undertake valuations of fixed assets on behalf of client companies which are subsequently reproduced, mentioned or referred to in public documents such as, Company Accounts, Stock Exchange Prospectuses and the like.

The Assets Valuation Standards Committee (AVSC), was set up in 1974 by the General Practice Divisional Council of the Institution. Since its formation it has published Guidance Notes, together with background papers for use by Members dealing with the valuation of company assets. The AVSC maintains close links with other professional bodies and Government Departments and promulgates advice to Members on the valuation of fixed assets based upon current trends and thinking. It monitors published valuations on behalf of the General Practice Division.

The AVSC is not concerned with the valuation figures themselves, merely with the *methods* employed by Members in arriving at those figures. Increasingly, over recent years, the Institution has become concerned that some valuations monitored by the AVSC have not been prepared in any consistent or logical form, with the result that persons reading published valuations may be misled by them.

The Extraordinary General Meeting of the Institution held in October 1990 approved the introduction of a new Bye-Law and Regulation, the text of which is set out in Appendix XVI. The Bye-Law will not take effect until it has been approved by Her Majesty's Privy Council. You will, therefore, need to check with the Institution's Professional Conduct Department on the date it becomes effective. If the Bye-Law is approved by the Extraordinary General Meeting, the Regulation will come into force on the same date as the Bye-Law.

A perusal of the Regulation and accompanying Notes will show that it is recognised that there may be exceptional circumstances which render the AVSC Practice Statements inappropriate. However, where a Member prepares a valuation which does not strictly comply with the Practice Statements he will be required to include in the valuation a statement to the effect that he has not followed the prescribed practice *and* give his reasons for not following the practice. It may be, for example, that the client has stipulated the basis on which the valuation is to be prepared. The Member must comply with the lawful instructions of his client (or decline instructions if he is not prepared to), but where he does depart from the Practice Statements the Report itself must make it quite clear that the prescribed Practice Statement has not been followed and give the reasons why it has not been followed.

21. Other Regulations

Some of the other Regulations relate only to estate agency:

- *Regulation 2* – Notification to the client vendor or lessor of terms and conditions, including conditions relating to charges and payment of expenses.
- *Regulation 3* – Prohibiting the stipulation to a purchaser or lessee when offering a property that the Member shall be retained.
- *Regulation 4* – Preparing contracts for the sale or letting of land for gain or reward.

But most of them, including, for example the following, apply to all activities:

- *Regulations 15 and 16* – Advertising and publicity – see Section 5, Chapter One.
- *Regulation 17* – References to the Institution – see Section 6, Chapter One.
- *Regulation 22* – Trade discounts.
- *Regulation 23* – Contraventions procedure.

Chapter Five

STEPPING OVER THE LINE

I hope the following will be of only academic interest to my readers. But should you find yourself faced with an enquiry from the Institution's Professional Conduct Department, an understanding of the disciplinary powers and procedures may be helpful.

22. The Institution's disciplinary powers

The Institution exercises its disciplinary powers through the Professional Conduct Committee, the Disciplinary Board, the Appeals Board and the General Council. The powers of the Professional Conduct Committee are limited. If it finds a Member in breach of the Institution's Bye-Laws or Regulations its powers are limited to:

(i) reprimanding or severely reprimanding the Member; and/or

(ii) requiring the Member to give an undertaking not to continue or repeat the conduct complained of or alleged; or

(iii) refering the subject matter of the complaint or allegation to the Disciplinary Board as a formal charge, although it does have the power to summarily suspend membership.

The Disciplinary Board has greater powers. It may, on finding a breach:

(i) reprimand or severely reprimand the Member;

(ii) require the Member to give an undertaking to refrain from continuing or repeating the conduct which is found to have constituted the contravention;
(iii) defer sentence for up to six months;
(iv) suspend the Member from membership for such period as it may determine;
(v) expel the Member from the Institution.

23. Criminal convictions, bankruptcy, insolvency and winding up

Bye-Law 25(2) gives the General Council of the Institution itself power to expel a Member summarily or to refer a matter to the Disciplinary Board. This power is not used lightly. It can only be used where:

(a) a Member has been convicted of a criminal offence involving 'embezzlement, theft, corruption, fraud or dishonesty of any kind or any other criminal offence carrying on first conviction a maximum sentence of not less than twelve months imprisonment;' or

(b) a notice has appeared in the London Gazette that a Member has been adjudicated bankrupt or there is produced a 'certified copy of a Deed of Arrangement he has entered into with or for the benefit of his creditors;' and this applies to Members in Scotland who may enter into a Deed of Arrangement for the benefit of their creditors or whose estates have been sequestrated; or

(c) a Member is a director of a Company which 'has had a winding-up order made in respect of it or has passed a resolution for voluntary winding-up (not being a member's voluntary winding-up)'; or

(d) the General Council receives satisfactory evidence, 'that the Member or a company carrying on practice as surveyors of which the Member is a director has become insolvent.'

In any such case the General Council may expel a Member forthwith or refer the matter to the Disciplinary Board for hearing.

They may also temporarily suspend a Member from membership pending a hearing by the Disciplinary Board.

Apart from its power under Bye-Law 25(2), the General Council also has powers to 'erase his name from the register, declare that he be no longer a Member and demand the surrender of his diploma'. It may exercise these powers if a Member fails to:

(1) pay money due from him to the Institution within three months of the money falling due to be paid; or
(2) pay any instalment of his subscription if he has elected to pay by way of instalments under Bye-Law 28(1); or
(3) deliver any certificate, report or other document he is required to deliver to the Institution within three months of the date by when it is due to be delivered.

These powers come from Bye-Law 31 and are used as the ultimate sanction when, for example, a Member fails to pay his subscription or fails to make a return under either the Members' Accounts Regulations or Compulsory Professional Indemnity Insurance Regulations.

24. The Professional Conduct Committee, Disciplinary Board and Appeals Board

The full text of Bye-Law 26 is set out in Appendix XIII. This guide is not intended to be an authoritative work on the Institution's disciplinary powers and procedures and I have therefore refrained from commenting at length on Bye-Law 26. I hope, however, that the following paragraphs may be of interest to the general reader in giving some explanation of the work of the Professional Conduct Committee, the Disciplinary Boards and Appeals Board.

The Institution's Professional Conduct Committee (PCC) is serviced by the Professional Conduct Department. The latter is a small team of administrative staff who deal with a wide variety of complaints against Members. Most of the staff have legal training. During the course of a year the department deals with in excess of 1,300 complaints of varying gravity and complexity. They work in conjunction with the Standards and Practice Department who are also responsible for monitoring Members' compliance with the Members' Accounts Regulations, Accountant's Report Regulations, Compulsory Professional Indemnity Insurance Regulations, and Notification of Particulars requirements (Bye-Law 24 (9B)). The Standards and Practice Department is also responsible for reviewing the Institution's Bye-Laws and Regulations and progressing changes to them.

Upon receipt of a complaint the Professional Conduct Department will seek to establish:

(1) the precise nature of the complaint;
(2) the identity of the Member concerned; and
(3) whether, prima facie, a breach of the Bye-Laws or Regulations has occurred.

The member will normally be sent a copy of the letter of complaint at an early stage and be asked for his comments and observations. At that stage a great many complaints are resolved. Only where the Member is unable to answer the complaint satisfactorily is the complaint referred to the PCC. In cases where a preliminary view is sought to establish whether a complaint should be referred to the PCC the Department's staff will either refer the papers to an outside legal adviser (the Disciplinary Solicitor) or to the Monitoring and Investigation Committee (MIC). It will consider whether the Member was professionally justified in his dealings with this client. The MIC, however, has no judicial powers. If it considers that there is prima facie evidence of a breach the case will be referred to the Professional Conduct Committee.

The PCC do not (at the moment) have the power to investigate complaints alleging negligence against a Member. In such cases the complainant may be referred to the Chartered Surveyors Arbitration Scheme (see Section 26), or will be advised to seek legal advice.

If a complaint is referred to the PCC the Member will be advised of the time and date of the hearing and that he may attend personally with a solicitor and counsel if he wishes or may simply write with his own observations and comments for the Committee to consider. The vast majority of complaints referred to the PCC are disposed of at this stage. The Committee may:

(1) find that there was no breach and dismiss the matter; or

(2) find a breach.

The PCC have additional powers to require a Member to appear before it to explain his conduct and may also order that a Member produce documents.

Since January 1990 the PCC now has an additional power to suspend a Member, in certain limited circumstances. Bye-Law 26A provides:

> '(1) *Where the Professional Conduct Committee receive evidence that a Member has failed:*
>
> *(a) to deliver to the Institution any certficiate, report or other document required by the Bye-Laws and Regulations; or*
>
> *(b) to comply with any provision of the Regulations made pursuant to Bye-Law 24(9A)*
>
> *the Committee may, instead of dealing with him under the provisions of Bye-Law 26 or transmitting particulars of such failure to the General Council for the exercise of its powers under Bye-Law 31, as the case may be, adopt the following procedure in his case.*'

So new is this power that at the time of writing it has happily not been invoked. Its primary intention is to ensure that the Professional Conduct Committee can act quickly to safeguard the public where, for example, a Member has been late in making a return under the Members' Accounts Regulations or has failed to make a return under the Compulsory Professional Indemnity Insurance Regulations, or has failed to notify particulars of his practice, thereby preventing the Institution from properly administering the Members' Accounts Regulations and the Compulsory

Professional Indemnity Insurance Regulations. The Committee may alternatively refer the Member to the General Council in order that the General Council may order that the Member's name be erased from the Register of Members. The Bye-Law goes on to provide:

> '(2) *The Committee may suspend such Member either forthwith or from such date as they may specify.'*

A Member who is suspended loses his entitlement to use the chartered designation and loses his right to vote. He is still subject to the Institution's Bye-Laws and Regulations. The Committee may order the suspension with immediate effect (if the Member appears before the Committee) or at some future date. The reason for this is to enable the Institution to advise the Member in writing that from a date in the future he will be suspended from Membership.

The Bye-Law continues:

> '(3)(a) *If the Committee decide to suspend a Member pursuant to paragraph (2) of this Bye-Law they shall forthwith notify the Member concerned and shall review their decision within twenty-eight days of the order for suspension taking effect.*
>
> (b) *On such review the Member shall be entitled to make representations in writing for consideration by the Committee.'*

It is hoped that where a Member is suspended summarily by the PCC he will rectify his conduct before the Professional Conduct Committee considers the suspension, as it is bound to do within 28 days of the suspension coming into effect. The Member may make representations to the Committee when it reviews its decision. If a Member has not rectified his conduct then under Bye-Law 26A(4)(b), the Committee may either proceed with its normal hearing, under the provisions of Bye-Law 26, and where they find a breach impose one of the penalties referred to above, or alternatively, may simply refer the matter to the Disciplinary Board as a formal charge. In any event, the Committee may order the suspension to continue. If the matter is referred to a

Disciplinary Board as a formal charge the Professional Conduct Committee can revoke the order for suspension at any time up to the date of the Disciplinary Board Hearing if the Member has rectified his conduct, otherwise the decision whether or not to continue the suspension rests with the Disciplinary Board.

Panels of the Disciplinary Board are made up of those Members appointed by the President to a permanent list. There are never less than three Members, one of whom must belong to the same division as the Member appearing before it. The panel always sits with a legal assessor who must be a barrister or solicitor of at least ten years standing. The PCC is represented by the Disciplinary Solicitor. The Disciplinary Board has power to order publication of its findings and any penalty imposed.

There is only limited appeal against the decision of a Disciplinary Board Panel. There is no appeal against the finding of a breach of the Bye-Laws or Regulations. A Member may only appeal against the penalty imposed. The appeal is heard by the Appeals Board which may confirm that penalty, reduce it or increase it. Unlike either the PCC or the Disciplinary Board, the Appeals Board also has the power to award costs, for or against the Member.

25. Precautions against disciplinary action

Over the past ten years or so, consumers have become increasingly aware of their rights to complain about goods and services. Most complaints arise out of misunderstandings. The majority of them can be avoided. One of the most avoidable complaints made to the Institution is that the complainant's surveyor has either delayed in replying or failed to reply to correspondence. It is common courtesy to reply to correspondence. In many cases all that it calls for is a short acknowledgement. Where either delay or failure to reply to correspondence is persistent the complainant naturally feels aggrieved. Members should observe the common courtesy of at least acknowledging correspondence from clients, former clients and other professional advisers within a week of receipt.

Little sympathy can be extended to a Member who not only fails to reply to correspondence from a client, but also fails to reply to correspondence from the Institution about that failure. If you receive a letter from the Institution alleging that you have either delayed in replying or failed to reply to correspondence, you would do well to reply immediately to the complainant and send a copy to the Institution.

Another common source of referrals to the Professional Conduct Committee concerns the failure of Members to file their Certificates and Accountant's Reports on time under Regulation 11 of the Members' Accounts Regulations. You will recall that Accountant's Reports must be filed within six months of the end of the accounting year to which they relate. To ensure that the Accountant's Reports are filed in time, Members should impress upon their Accountants the importance of dealing with the audit quickly after the end of the accounting period. Inevitably, accountants will have queries to resolve and occasionally it may not be possible for the accountant to prepare the report before the end of the six month period. In such cases, the Member should ask his accountant to write to the Institution explaining the reason for the delay. This may not prevent the Member from being disciplined, but may give him a greater chance of leniency being shown to him by the Institution.

The Institution has recently installed a computerised monitoring system. Members who are required to submit returns will be contacted by the Institution if they are late in making the returns. Members who fail to submit returns on time will be reported to the Monitoring and Investigation Committee who may themselves refer the matter to the Professional Conduct Committee. A Member could now find that he has been suspended from Membership because of delay in filing his Accountant's Report under the PCC's new power.

Members may also find that they have been selected at random, or due to an apparent breach of the Regulations, for a spot check of their accounts by independent accountants appointed by the Institution. This is a necessary part of the monitoring system and is insisted upon by the insurers of the Clients' Money Protection Scheme. Members who are selected should make every effort to co-operate with the accountants, who will normally spend no more than one day at their offices.

Another common area of complaint concerns Members who find themselves in financial difficulty. In times of recession and high interest rates chartered surveyors are just as susceptible to cash flow difficulties as other professionals. The Royal Institution of Chartered Surveyors Benevolent Fund is sometimes able to provide assistance. The fund was created to help Members and their dependants and even if they are not able to provide temporary financial assistance, they may be able to help in other ways. It only requires a telephone call. (The Royal Institution of Chartered Surveyors Benevolent Fund Limited, 2nd Floor, Tavistock House North, Tavistock Square, London WC1H 9RJ, telephone 071 387 0578/9.)

26. Chartered Surveyors Arbitration Scheme

Though quite separate from the Institution's disciplinary function, a mention of this scheme is relevant in the context of responding to clients' complaints.

The scheme is intended to be an inexpensive and informal method of settling disputes between chartered surveyors and their clients where compensation is sought. The scheme does not deal with issues of dishonesty or other alleged professional misconduct. Before the scheme can operate, the Member concerned must give his consent. He, therefore, has the opportunity to consult his professional indemnity insurers on the appropriateness of the scheme to the matter at issue. The arbitration is conducted in writing. Further details of the scheme can be obtained from the Institution direct or from the Chartered Institute of Arbitrators which administers the scheme independently from the RICS. Its address is:

International Arbitration Centre, 75 Cannon Street, London EC4N 5BH, telephone 071 236 8764.

Appendix I

BYE-LAW 24: RULES OF CONDUCT (DISCIPLINE)

24. (1) No Member shall conduct himself in a manner unbefitting a Chartered Surveyor.
(2) No Member shall be connected with any occupation or business in any way which would, in the opinion of the General Council, prejudice his professional status or the reputation of the Institution.
(3) Subject to the Regulations, it shall be the duty of every Member:
(a) when acting for a client or when in contemplation of acting for a prospective client whose interests conflict or may conflict with his own or those of any of his associates as defined in the Regulations, to disclose the relevant facts forthwith to the client or prospective client and to the associate and where such disclosure is oral to confirm the same in writing at the earliest opportunity and inform the client that he will be unable to act or continue to act unless the client requests him so to do after obtaining independent professional advice;
(b) to ensure that neither he personally nor any firm or company carrying on practice as surveyors of which he is a partner or director acts for two or more parties with conflicting interests without disclosing the relevant facts to each of those parties forthwith and confirming such disclosure in writing at the earliest opportunity.
(4) No Member shall carry on practice as a surveyor through the medium of a company except in accordance with the Regulations.
(5) (a) For the purposes of this paragraph (5) of this Bye-Law
(i) 'Partner' shall include a sole principal of a practice or a partner in a firm or a Director of a company;
(ii) 'firm' shall mean a practice, firm or company carrying on practice as surveyors;
(iii) 'contravention' shall include any act or omission which would, if committed by a Member, constitute a contravention.

(b) Every Member who
   (i) is or holds himself out or allows himself to be held out to be a Partner in a firm; or
   (ii) allows his name and/or designatory letters or designation to appear on the notepaper or in the advertisements of any firm in which no Partner is a Member; or
   (iii) is a Partner in a firm which is so connected with another firm, in which no Partner is a Member, that in the opinion of the General Council persons dealing with that other firm believe or may be induced to believe that the two firms are connected;

shall be held responsible for any contravention of the Bye-Laws or the Regulations committed by any Partner or by any member of the staff of such firm or other firm, provided that if such Member shall show that without any default on his part he had no reason to be aware and was not aware of any such contravention and that he had prior to such contravention taken all such steps as may be reasonable to ensure that such contravention was not committed he may be acquitted of responsibility for such contravention.

(6) Subject to paragraph (10) of this Bye-Law and the Regulations no Member shall:
   (a) directly or indirectly exert undue pressure or influence on any person, whether by the offer or provision of any payment, gift or favour or otherwise, for the purpose of securing instructions for work, or accept instructions from any person on whom he has reason to believe that undue pressure or influence may have been exerted by a third party in expectation of receiving a reward for the introduction;
   (b) quote a fee for professional services without having received sufficient information to enable the Member to assess the nature and scope of the services required;
   (c) having once quoted a fee for professional services revise that quotation to take account of the fee quoted by another member of the surveying profession for the same services; or
   (d) quote a fee for professional services which is to be calculated by reference to the fee quoted or charged by another member of the surveying profession reduced by some proportion or amount.

(7) No Member shall invite instructions for work except in accordance with the Regulations.

(8) Subject to the Regulations every Member shall:
   (a) keep in one or more bank accounts separate from his own, his firm's or his company's bank account (as the case may be)

any clients' money held by or entrusted to him, his firm or his company in any capacity other than that of beneficial owner;

(b) account at the due time for all moneys held, paid or received on behalf of or from any person (whether a client or not) entitled to such account and whether or not after the taking of such account any payment is due to such person; and

(c) keep such accounting records as are specified in the Regulations and maintain them in accordance with the Regulations.

(9) No Member shall carry on practice as a surveyor under any such name, style or title as to prejudice his professional status or the reputation of the Institution.

(9A) Every Member shall, in accordance with the Regulations, be insured against claims for breach of professional duty as a surveyor.

(9B) Every Member shall, in accordance with the Regulations, furnish to the Institution such particulars of his practice, employment and business as it may reasonably require for the administration of the Institution and for the regulation of Members' professional conduct and discipline.

(10) Nothing in this Bye-Law or the Regulations shall be construed so as to restrict or to permit the restriction of any Member acting as an estate agent in England or Wales with respect to the matters contained in the Restriction on Agreements (Estate Agents) Order 1970.

Appendix II

BYE-LAW 5: DESIGNATIONS

5. (1) The designation of Members by distinguishing initials or words shall be as follows:
Every Fellow shall be entitled to use after his name the initials FRICS (i.e. Fellow of The Royal Institution of Chartered Surveyors) and the designation 'Chartered Surveyor'. Every Professional Associate shall be entitled to use after his name the initials ARICS (i.e. Professional Associate of The Royal Institution of Chartered Surveyors) and provided that the fact that he is a Professional Associate is disclosed the designation 'Chartered Surveyor'.

(2) (a) A firm whose partners consist entirely of Fellows and Professional Associates may use in conjunction with its title the designation 'Chartered Surveyors';

(b) a company whose directors consist entirely of Fellows and Professional Associates may use in conjunction with its corporate name (but not with its trading name if different from its corporate name) the designation 'Chartered Surveyors';

provided that each individual partner of that firm or director of that company who is a Professional Associate shall insert after his name where shown the initials appropriate to his class of membership.

(3) (a) A Fellow and (subject to the proviso set out in paragraph (1) of this Bye-Law) a Professional Associate who is a member of a Division listed in the first column of the Table below shall be entitled to use the alternative designation appearing against that Division in the second column of the Table.

(b) A firm whose partners or a company whose directors consist entirely of Fellows and Professional Associates may if all the partners or all the directors are entitled to one of the individual alternative designations listed in the second column of the Table below use in conjunction with its title (in the case of a firm) or its corporate name (in the case of a company) the alternative designation appearing against the individual alternative designation in the third column of the Table, subject to the proviso set out in paragraph (2) of this Bye-Law.

	Alternative Designation	
Division	Individual	Firm
Quantity Surveyors	Chartered Quantity Surveyor	Chartered Quantity Surveyors
Land Surveyors	Chartered Land Surveyor	Chartered Land Surveyors
Building Surveyors	Chartered Building Surveyor	Chartered Building Surveyors
Minerals	Chartered Minerals Surveyor	Chartered Minerals Surveyors
General Practice Planning and Development Rural Practice	Chartered Valuation Surveyor	Chartered Valuation Surveyors

(4) A Member of the General Practice Division or of the Rural Practice Division who is not registered on the Land Agency Register and who is entitled to use the alternative designation 'Chartered Valuation Surveyor' shall be further entitled to add to that designation the words 'and Estate Agent'. A firm all of whose partners or a company all of whose directors are so entitled may add to the alternative designation 'Chartered Valuation Surveyors' the words 'and Estate Agents', subject to the proviso set out in paragraph (2) of this Bye-Law.

(5) A member of the Rural Practice Division who is registered on the Land Agency Register and who is entitled to use the alternative designation 'Chartered Valuation Surveyor' shall be further entitled to add to that designation the words 'and Land Agent'. A firm all of whose partners or a company all of whose directors are so entitled may add to the alternative designation 'Chartered Valuation Surveyors' the words 'and Land Agents', subject to the proviso set out in paragraph (2) of this Bye-Law.

(6) Members shall not use in conjunction with any of the above designations any other designation in such a way as to imply that such other designation is also a Chartered designation.

(7) Members shall not
  (a) use the designatory letters 'FLAS' or 'QALAS' or 'FAI' or 'AAI' or
  (b) associate with or in any way support any body having as a title 'The Chartered Land Agents' Society' or 'The Chartered Auctioneers' and Estate Agents' Institute' or any similar title.

(8) For the purposes of this Bye-Law 'company' shall mean an unlimited company or a company limited by shares or guarantee through which Members may be permitted to practise pursuant to the Bye-Laws and Regulations.

REGULATIONS MADE UNDER BYE-LAW 24: CONDUCT REGULATIONS

Definitions

1. In these Regulations the following expressions shall unless the context otherwise requires have the meanings respectively assigned to them, namely:
   'Member' includes a Member practising as a surveyor or a firm or company carrying on practice as surveyors in which at least one partner or one director, as the case may be, is a Member.
   The expressions 'lessor' and 'lessee' in these Regulations shall be deemed to include, in relation to the grant of a right to extract minerals, the grantor and grantee, respectively, of such a right; and references to letting shall be construed accordingly.
   The word 'company' in these Regulations shall mean a company limited by shares or an unlimited company, wherever incorporated.
   'Estate agency services' includes the sale, letting or acquisition (by purchase or lease) of landed property on behalf of a client.
   'client' means a person or body who retains a Member and is responsible for the payment of the Member's fee or commission.

Notification of Charges

2. A Member acting for a vendor or lessor in the sale or letting of property shall, when accepting instructions, notify his client of the terms and conditions, including conditions relating to his charges and the payment of expenses, on which he is to act; provided that this Regulation shall not apply in a case where the client is already aware of such terms and conditions as aforesaid.

Stipulations

3. A Member shall not be permitted, when offering property to a prospective purchaser or lessee on behalf of a client, to stipulate that he shall be retained by the purchaser in any capacity.

Contracts for Sale or Letting

4. Save as regards tenancies for terms not exceeding three years a Member (unless otherwise professionally qualified to do so) shall not be permitted:
   (a) except in the case of sales by auction to draw up or prepare for fee, gain or reward any contract or document purporting to be a contract for the sale or letting of land or buildings;
   (b) except in the case of sales by auction or tender, to present to any person for signature a contract or document purporting to be a contract for the sale or letting of land or buildings without pointing out to that person the desirability of taking legal advice.

   Provided that paragraph (a) of this Regulation shall not apply to Members practising in Scotland, Northern Ireland or the Republic of Ireland.

Site Boards

5. (1) Except when acting as an estate agent, a Member shall not be permitted to exhibit his name, or the name of his firm or company, as the case may be, on a building site otherwise than by means of the Institution's standard name board, or a similarly unostentatious board on which the lettering does not exceed such height as may be permitted on the Institution's standard name board.
   (2) For the purposes of this Regulation, the Institution's standard name board shall be a board manufactured in accordance with such specification as may from time to time be promulgated by the Institution.
   (3) The Institution's standard name board may be used only:
      (a) to exhibit the name and address of a firm or company, as the case may be, entitled to use a designation permitted under Bye-Law 5; or
      (b) to exhibit the name and position of a Member employed by a public authority in respect of works carried out by that authority under his direct supervision, and where he is the senior officer concerned with such works.

Connected Businesses

6. Without prejudice to the generality of Bye-Law 24(2) a Member carrying on practice as a surveyor:
   (a) shall not be permitted to take part directly in the management or control of any organisation whose business consists wholly or substantially, or in the case of a Member performing the function of quantity surveying, wholly or partly, of building or civil engineering contracting;

(b) shall be permitted to conduct, or allow to be conducted, as an adjunct to or in conjunction with his practice as a surveyor any business (a 'connected business') which in the opinion of the General Council is of such a nature that neither the Member's professional status nor the reputation of the Institution is thereby prejudiced.

(c) In any case where the General Council is satisfied that a connected business is of such a nature or has been conducted in such a manner as to prejudice a Member's professional status or the reputation of the Institution, it may:

(i) prohibit that Member from conducting or allowing to be conducted such business, or any similar business, as a connected business and/or

(ii) deal with that Member in any one or more of the ways specified in Bye-Law 25(1).

7. Every Member who is employed in a company or undertaking not primarily engaged in the provision of surveying services, and who in the course of such employment offers or provides surveying services to third parties, shall at the earliest opportunity make clear to the recipient or prospective recipient of such services:

(a) the identity of the Member's employer; and

(b) the nature of any business in which the Member's employer engages.

Conflict of Interest

8. (1) Without prejudice to the generality of Bye-Law 24(3) where a Member is instructed:

(a) to sell or let a property and it is intended that the property be acquired by the Member himself or an associate of his, or

(b) to negotiate the purchase or lease of a property owned by the Member himself or an associate of his

he must

(i) make full and immediate disclosure to his client of all the relevant facts;

(ii) make a declaration of his interest to the vendor's or lessor's, or purchaser's or lessee's, solicitors, as the case may be;

(iii) inform the client that he will be unable to act or continue to act unless the client requests him so to do after obtaining independent professional advice; and

(iv) confirm the requirements of (i), (ii) and (iii) above in writing at the earliest opportunity.

(2) Where a Member acts as agent for the sale or letting of property owned by the Member himself or an associate of his, or in which he

or an associate of his has an interest, he must disclose the relevant facts to the prospective purchaser or lessee or their solicitors.

(3) Where a Member conducts a connected business as permitted by Regulation 6(b), or where he allows such a business to be conducted, and where the interests of a client of the practice conflict or may conflict with the interests of a client or customer of the connected business, he must disclose the relevant facts forthwith to all parties concerned and confirm the disclosure in writing at the earliest opportunity.

(4) In Bye-Law 24(3) and this Regulation 'associate' may include the Member's firm or company or a partner of his in the case of a firm or co-director in the case of a company, his spouse, a near relative of his by blood, adoption or marriage, his nominee or employee, or any other person, firm or company so associated with the Member that his or its interest may conflict with that of the client.

Financial Interest

9. (1) No member shall take a financial interest in any matter in which he acts in his practice as a surveyor nor shall he accept or offer to accept instructions on terms which could be construed as taking a financial interest provided:
   (a) that a fee or commission at the rate usually charged by the Member for similar work; or
   (b) an escalating commission rate, taking into account the value of the work to the client, which has been expressly agreed in writing with the client

   shall not be construed as taking a financial interest.

   (2) A Member performing the function of quantity surveying shall not, without disclosure to all the parties, permit himself to be named in the contract or elsewhere if he is associated with one of the parties in such a way that he may gain or lose financially, apart from his normal professional fees, according to the amount at which the contract is settled.

Incorporation with Limited or Unlimited Liability

10. (1) A Member may carry on practice as a surveyor through the medium of a company provided that he complies with these Regulations.

   (2) No Member shall carry on practice as a surveyor as aforesaid in any case where he is a director of the company concerned unless:
   (a) a provision to the following effect is included in the Memorandum of Association or equivalent constitutional document of that company and in such manner that it

qualifies those powers of that company enabling it to offer surveying services:

'Any business of surveying for the time being carried on by the company shall at all times be conducted in accordance with the Rules of Conduct for the time being of The Royal Institution of Chartered Surveyors';

(b) a provision to the following effect is included in the Articles of Association or equivalent constitutional document of that company:

'It shall be the duty of the Directors to ensure that any business of surveying for the time being carried on by the company shall at all times be conducted in accordance with the Rules of Conduct for the time being of The Royal Institution of Chartered Surveyors.'

11. Notwithstanding the provisions of Regulation 10, a Member who is a livestock auctioneer or a member of the Land Surveyors Division to whom the Institution has, prior to 21 October 1986, granted dispensation to practise through the medium of a limited company, shall be permitted to practise as aforesaid in accordance with the terms of that dispensation.

12. (1) The restrictions imposed by Regulation 10 shall only apply in the case of a company which undertakes or seeks work in any one or more of the United Kingdom, the Republic of Ireland, the Channel Islands, the Isle of Man and Hong Kong.

(2) No Member shall carry on practice as a surveyor as aforesaid in any country other than the United Kingdom, the Republic of Ireland, the Channel Islands, the Isle of Man or Hong Kong if to do so would conflict with the laws of that country or with the rules of the relevant professional society, if any, in that country.

Status and Designations

13. In any list of partners and/or staff in a firm, or directors and/or staff of a company, carrying on practice as surveyors, published by or on behalf of a Member or such firm or company, which includes the names of one or more Members, it shall be the duty of every Member or Members so named to ensure that his or their status within the firm or company is clearly stated and that no chartered designation is used in such a way as to give the impression that the firm or company is entitled to use that designation if that is not the case.

Inducements for the introduction of clients

14. Without prejudice to the generality of Bye-Law 24(6)(a) a Member shall be permitted to pay a fee or commission, or provide a gift

or favour, to a third party in recognition of the introduction of a client, if:

(a) he has disclosed in writing to the prospective client before accepting instructions the amount or nature, as the case may be, of the fee, commission, gift or favour, and the identity of the third party, (except that such disclosure will not be required where the introduction has been effected in the ordinary course and/or by reason of the third party's surveying business); and

(b) he has no reason to believe, and does not believe, that undue pressure or influence was exerted on the prospective client by the third party.

Advertising in association with non-surveying undertakings

15. No Member shall publicise his services, or allow his services to be publicised, in association with any goods or services available, from any other source in such a way as, in the opinion of the General Council, might call into question the independence of his professional advice or give rise to a conflict of interest.

Professionalism, Accuracy and Clarity

16. Every Member shall ensure that any publicity for which he may be held responsible is neither inaccurate nor misleading nor likely to cause public offence.

References to the Institution

17. No Member shall:

(a) purport to represent the views of the Institution unless expressly authorised so to do; or

(b) publicise the Institution or its Members generally in terminology which has not either already appeared in an advertisement published by the Institution or received the approval of the Institution.

Avoidance of claims of Superiority

18. No Member shall make a public statement claiming an advantage or superiority over other firms, unless such a statement can be substantiated to the satisfaction of the General Council.

Confidentiality of Clients' Affairs

19. (i) No Member shall, without his client's consent, disclose personal information concerning that client.

(ii) For the purpose of this Regulation 'personal information' shall include any information relating to the personal, financial or business circumstances of any person or body, or any other information of a sensitive nature, which, having been disclosed to the Member in his professional capacity is not known to have been otherwise generally published.

Inviting Instructions

20. A Member may invite instructions for work provided that he complies with paragraphs (1) to (3) of this Regulation:
    (1) No member may invite instructions for work
        (a) by personal call to a private address; or
        (b) by telephone call to a private address except in response to an advertisement of a property for sale or to let.
    (2) Where a Member invites instructions for estate agency services in any communication which in the opinion of the General Council constitutes a personal approach, such communication shall include a clear statement that if the prospective client has already appointed a sole agent, a liability to pay two commissions may arise if a second agent is appointed otherwise than as a sub-agent to that sole agent; and a clear statement of the circumstances in which commission will be required; and where such statements are oral they shall be confirmed in writing at the earliest opportunity.
    (3) Where a Member invites instructions for professional services other than estate agency services in any communication as aforesaid, such communication shall include clear statements that
        (a) the Member will not accept instructions for work currently in hand and for which another professional adviser has already been retained until the Member has first satisfied himself that the previous instructions have been terminated; and
        (b) the prospective client could, if a professional adviser were appointed to succeed the adviser already retained, be liable to pay fees to both professional advisers;

    and where such statements are oral they shall be confirmed in writing at the earliest opportunity.

Clients' Money: Market Accounts

21. Notwithstanding Bye-Law 24(8) and the Members' Accounts Regulations a Member who is a livestock auctioneer or a chattel auctioneer may keep moneys belonging to himself, his firm or his company (as the case may be) in the same bank account

as moneys held on behalf of clients provided such mixed account is:

(a) appropriately identified; and
(b) operated exclusively for purposes connected with the conduct of the Members' livestock market (and other live or dead stock sales) or chattel sale-room business.

Trade Discounts

22. (1) No Member shall in his professional capacity accept otherwise than for the benefit of his client any trade or other commercial discount or commission from any trader in respect of any item of agricultural machinery or equipment, or from any trader whose business consists in the provision of goods or services used in the building construction or building maintenance industries.
(2) Subject to sub-paragraph (1) hereof a Member may accept a discount in respect of goods or services ordered by him on behalf of a client provided full disclosure is made to the client.

Contraventions Procedure

23. Where it appears that a contravention of the Regulations pertaining to accuracy and clarity, claims of superiority and inviting instructions may have been committed by a Member, no steps shall be taken by the Institution to investigate a complaint made by another Member unless either:
(a) (i) the complainant shall have notified the Member concerned of his complaint direct or through the Chairman or Honorary Secretary of the Branch, or of the Division of the Branch, to which the Member concerned belongs and invited the Member's comments thereon; and
(ii) the Member shall have failed to satisfy the complainant that there are no sufficient grounds for taking any further action; or
(b) the complainant shall have satisfied the General Council that there are sufficient grounds for steps to be taken by the Institution without the procedure set out above having been adopted.

The procedure set out in this Regulation is without prejudice to the right of the Institution at its discretion to take appropriate steps in the absence of a complaint or on a complaint by a member of the public.

Notification of Practice Details

24. In pursuance of his obligation under Bye-Law 24(9B)
(a) every Member shall within 28 days of being required to do so

furnish to the Institution such particulars in such form as the General Council may reasonably require

(i) of his practice, if he is carrying on professional practice as a sole principal, partner or director; and

(ii) of his employment if he is employed under a contract of service or a contract for services;

(b) where a Member has furnished particulars in accordance with paragraph (a) of this regulation, and where any change occurs in the circumstances notified in those particulars, he shall furnish full particulars thereof to the Institution no later than 7 days after such change has come into effect.

BYE-LAW 9(3) AND THE CONTINUING PROFESSIONAL DEVELOPMENT REGULATIONS

Bye-Law 9(3) provides as follows:

'Every Professional Associate and Fellow of the Institution shall for so long as he remains a Member undergo in each year such continuing professional development and shall from time to time provide to the Institution such evidence that he has done so as the Regulations shall provide.'

Continuing professional development regulations

Made Pursuant to Article 18 of the Supplemental Charter

1. Continuing Professional Development is the systematic maintenance, improvement and broadening of knowledge and skill and the development of personal qualities necessary for the execution of professional and technical duties throughout the practitioner's working life. Every Member of the Institution as specified in Bye-Law 9(3) must comply with these Regulations. Other Members are invited to observe them as part of their professional responsibility.
   (*Note*: The Regulations only apply to Professional Associates elected on or after 1 January 1981. As from 1 January 1991 it applies to all Fellows and Professional Associates.
2. Members themselves are responsible for keeping a record of their participation in qualifying activities, including (where relevant) the date(s), subject matter, speakers and total time computed in accordance with Regulation 5. They must keep the record for three years after the qualifying activity and send a copy to the Institution when so requested.

3. Continuing Professional Development may take the following forms:
   (a) courses and technical meetings organised by:
      (i) the Institution, its Branches, or Institution and College Conferences;
      (ii) universities, polytechnics or other colleges;
      (iii) employers of chartered surveyors;
      (iv) other professional bodies;
      (v) the College of Estate Management; and
      (vi) other relevant course-providers;
   (*Note*: The Institution does not approve courses or their content, but course organisers may issue certificates of attendance to chartered surveyors who request them; such certificates should include the details listed in Regulation 2 above).
   (b) discussion meetings on technical topics;
   (*Note*: there should in this instance be a competent person in charge of the proceedings, the subject should be announced in advance and the meetings should have some formalised structure including for example an introductory paper, video or tape).
   (c) private studies of a structured nature on specified themes;
   (*Notes:* (i) prescribed pre-course reading may be counted as private study;
   (ii) regular reading of professional journals may not normally be counted as private study).
   (d) correspondence courses, Open University courses, or other supervised study packages, being a programme of reading or recorded lectures;
   (e) research or post-qualification studies; or
   (f) authorship of published technical work or the time spent in preparation and delivery of lectures in connection with a qualifying CPD event or other similar professional or technical meetings.
4. To qualify as Continuing Professional Development, the activities described in Regulation 3 must be related to:
   (a) some part of the theory and practice of surveying as defined in the Royal Charter; and/or
   (b) other technical topics related to a member's current or potential occupation; and/or
   (c) personal or business skills designed to increase a member's management or business efficiency.
5. Members specified as in Bye-Law 9(3) and Regulation 1 must complete 60 hours Continuing Professional Development in every consecutive period of three years, computed as follows:
   (a) the time attributable to any event shall be the duration from the formal opening to the formal closing, calculated to the nearest half-hour;

(*Note:* when an activity extends over more than one day, only the formal teaching sessions may be counted);

(b) when a member attends for only part of an activity only the time attended shall be counted;

(c) time recorded under Regulation 3(c) can constitute no more than two thirds of the total requirement;

(d) time spent in administering a qualifying activity shall not be counted.

6. Members who have completely retired from practice are exempted from these requirements.

Appendix V

COMPULSORY PROFESSIONAL INDEMNITY INSURANCE REGULATIONS MADE PURSUANT TO BYE-LAW 24(9A)

Regulations

Definitions

1. In these Regulations unless the context otherwise requires:
   'Member' shall mean a Member who is or who is held out to the public to be practising as a surveyor as:
   (a) a sole principal or a partner or a director of, as the case may be, a practice, firm or company carrying on practice as surveyors; or
   (b) a consultant to a practice, firm or company carrying on practice as surveyors; or
   (c) a Member who has ceased after 1 January 1986 to carry on practice as a surveyor in any of the capacities listed in sub-paragraphs (a) and (b) above and is not otherwise covered by insurance in accordance with these Regulations for work carried out during the period when he was in practice in any of those capacities.

   'Sole principal' shall include a Member who carries on practice as a principal in addition to other employment.
   'Company' shall mean a company incorporated with unlimited or limited liability.
   'Firm' shall include a partnership.
   'Consultant' shall include any Member whether or not expressly described as a consultant whose name and/or designatory letters or designation appears on the notepaper or in the advertisements of any practice, firm or company carrying on practice as surveyors in which no director or partner is a Member.

Scope of Cover Required

2. (a) Subject to sub-paragraphs (b) and (c) of this Regulation 2, every Member shall procure that any practice, firm or company carrying on practice as surveyors of which he is a principal, partner or director, as the case may be, shall be insured so that such practice, firm or company and its principal, partners or directors, as the case may be, are covered against claims arising out of work undertaken or performed within the United Kingdom (including the Channel Islands and the Isle of Man) and/or within the Republic of Ireland for breach of professional duty as a surveyor. Provided always that if a Member practising solely in the capacity mentioned in Regulation 1(b) can show:
   (i) that the practice, firm or company to which he is a consultant is insured; and
   (ii) that such policy of insurance names the Member as the insured or one of the insured; and
   (iii) that such policy of insurance gives the same cover to the Member as required by these Regulations

   then that Member shall not be under an obligation to carry any separate insurance over and above that carried by the practice, firm or company concerned.

(b) Members shall insure by means of a policy no less comprehensive than the form of the Professional Indemnity Collective Policy as issued by RICS Insurance Services Limited (in so far as this relates to cover against breach of professional duty as a surveyor as set out in the preceding sub-paragraph and with the exception of the Institution's special conditions) at the time when any policy of insurance is taken out pursuant to these Regulations.

(c) The minimum amount of cover required under these Regulations shall be:
   (i) £100,000 for each and every claim where the gross income of the practice, firm or company in the preceding year did not exceed that amount; or
   (ii) £250,000 for each and every claim where the gross income of the practice, firm or company in the preceding year exceeded £100,000.

   For the purpose of these Regulations the expression 'gross income' shall include all professional fees, remuneration, commission and income of any sort whatsoever in so far as these have been derived from work undertaken or performed in the United Kingdom (including the Channel Islands and the Isle of Man) and/or within the Republic of Ireland but excluding any sums received for the reimbursement of disbursements, any amount charged by way of

value added tax and any income from judicial or other such offices as the General Council may from time to time determine.
For the purposes of these Regulations 'preceding year' shall mean the Member's accounting year which ended during the twelve months before the date on which any insurance policy under these Regulations is taken out.

(d) Where a practice, firm or company merges with or takes over or succeeds to the whole or any part of any other practice, firm or company then the gross income for the purposes of these Regulations of the new or merged or successor practice, firm or company shall be the total of the gross fees of the practices, firms or companies which were merged or taken over during the preceding year.

Minimum Amounts

3. The minimum amounts of cover specified in Regulation 2 of these Regulations shall apply:
   (a) to a practice, firm or company irrespective of the number of partners in the firm or directors of the company;
   (b) to each individual firm, where there is more than one, with which a Member may be involved in any of the capacities listed in sub-paragraphs (a) and (b) in the definition of Member in Regulation 1 of these Regulations.
4. (a) In the case of a Member who carries on practice as a principal in addition to other employment the gross income for the purposes of Regulation 2 shall be the gross income as defined above derived by the Member from practice as a principal.
   (b) In the case of a Member in category (c) of the definition of Member in Regulation 1 above the gross income for the purposes of these Regulations shall be the gross income derived from the last full year of practice as a surveyor.

Uninsured Excess

5. The amount of any claim which a Member or the relevant practice, firm or company may be required to pay before any indemnity is granted under the terms of any policy of insurance required under these Regulations ('the uninsured excess') shall not exceed an amount equal to 2.5 per cent of the minimum cover required for such Member, practice, firm or company under these Regulations; provided always that if a Member, practice, firm or company is insured for an amount of cover greater than the minimum amount required under these Regulations the uninsured excess under such insurance may be for any amount up to a maximum of 2.5 per cent of the full limit of indemnity of such insurance.

Monitoring

6. Members shall provide to the Institution within 28 days of being required to do so such evidence and in such form as the General Council may from time to time prescribe either that the Member is not subject to these Regulations or that the Member has complied therewith.

Waiver

7. The General Council shall have power to waive in writing in a particular case any of the provisions of these Regulations.

Commencement

8. These Regulations shall come into force on 1 January 1986.

Notes to the Regulations

The Regulations apply to Members, as defined in Regulation 1, only in respect of work undertaken by them or by their practice, firm or company within the United Kingdom (including the Channel Islands and the Isle of Man) and/or the Republic of Ireland.

(i) Members as defined in Regulation 1 who currently undertake no work in the United Kingdom and/or the Republic of Ireland are reminded that they are required under the Regulations to effect run-off cover in respect of any such work undertaken by them in the past.

(ii) Although the Regulations do not require Members to insure in respect of work which is undertaken outside the United Kingdom and/or the Republic of Ireland, the General Council reaffirms its commitment to the principle that Members should obtain the best available cover for all work that they undertake wherever it is undertaken. Accordingly the General Council expects Members to abide by the spirit of the Regulations in so far as they are compatible with the laws of the countries in which they undertake work.

(iii) Members are reminded that exposure to liability may be determined by factors other than where the work is undertaken. Thus, work undertaken within the United Kingdom and/or the Republic of Ireland may also give rise to liability under foreign jurisdiction and conversely work undertaken overseas may give rise to a cause of action within the Courts of the United Kingdom and/or the Republic of Ireland. If Members are in any doubt as to the jurisdictions within which actions could be brought against them in respect of any work contemplated, they should seek legal advice before accepting instructions for such work, and effect insurance accordingly.

(iv) RICS Insurance Services Limited (see note (vii) below) have wide experience of obtaining professional indemnity insurance cover overseas and Members seeking such cover are recommended to contact that company in the first instance.

(v) Members should note that professional advice for which no fee is charged may give rise to a claim.

(vi) Members who are employees are advised to ensure that they are indemnified by their employers against claims arising from alleged professional negligence in the course of their employment.

(vii) The Royal Institution of Chartered Surveyors' Professional Indemnity Collective Policy is issued by RICS Insurance Services Limited through which company Members are recommended to insure in order to be satisfied that their policy complies with the requirements.

(viii) Members are further recommended to give serious consideration to obtaining public liability insurance. Such insurance is concerned with liabilities to third parties for injury to their persons or property which fall outside the scope of professional indemnity insurance. Claims of that nature can be very substantial. The RICS Public Liability Policy, obtainable through RICS Insurance Services Limited, enables Members to obtain indemnity against that risk and relieves them of the need to take personal injury liability into account when determining the sum insured in their professional indemnity insurance policy.

(ix) Members are reminded that if a claim is made against them the insurance must be sufficient to meet not only the damages awarded, but also interest on such damages and the costs of the person making the claim. Interest and costs can significantly increase the amount payable in respect of claims. The amounts specified in the Regulations are *minimum* amounts, and Members may need much higher levels of cover. Members are strongly urged to take advice from an insurance broker or other properly qualified person as to the proper level of cover needed for any Member or practice.

(x) A copy of the Professional Indemnity Collective Policy issued by RICS Insurance Services Limited is freely available from the Institution. Members are strongly advised to obtain a copy before each renewal of insurance as the exclusions may be altered from time to time.

Appendix VI

MEMBERS' ACCOUNTS REGULATIONS

Part 1 – Definitions

Regulation 1

In these Regulations unless the context otherwise requires:

'clients' money' shall mean money held or received by a Member, his firm or his Company on account of a person for whom he, his firm or his company is acting either as a surveyor or, in connection with his practice as a surveyor, as agent, bailee or in any other capacity including that of stakeholder provided that the expression 'clients' money' shall not include money to which the only person beneficially entitled is the Member himself, or money held in an account by a Member jointly with a third party (not being a client) and over which the Member does not have a power of withdrawal on his sole signature or the signature of himself or any partner, co-director, servant, employee or other person connected with his firm.

'client' shall mean any person or body on whose account a Member holds or receives clients' money.

'client account' shall mean a current or deposit account in the name of the Member at a bank in the title of which account the word client appears or a separate account opened in the name of some person designated in writing by a client over which the Member has power of withdrawal on his sole signature or the signature of himself or any partner, co-director, servant, employee, or other person connected with his firm.

'bank' shall mean the Bank of England, a branch or wholly owned subsidiary of such other organisations as are banking or discount companies on the list maintained for the purposes of the Banking Act 1979, a Trustee Savings Bank within the meaning of Section 3 of the Trustee Savings Bank Act 1969, the Post Office and certain Building Societies as defined in note (i) to this definition.

'Member' shall include a Member who is a sole principal of a practice or a partner in a firm or a director of a company or is held out to the public as a partner in a firm or director of a company carrying on practice as a surveyor.
'firm' shall mean a partnership.
'company' shall mean either an unlimited or a limited liability company.

Notes to Regulation 1

'clients' money'

(i) It should be appreciated that specific instructions of the client (in writing and acknowledged by the Member in writing) take precedence over the Regulations in relation to money received or paid on his behalf (e.g. Regulation 8(2)(c)). For example a client may instruct that his money be withheld from a client account, retained in the Member's Office in the form of cash, or banked in any account which is not a client account, e.g. a special, named, management account. He may give instructions as to payments of money held on his behalf in any manner he wishes even though the Member may strongly advise against such payment.

(ii) A 'float' put in by a Member to keep his client account in balance is not clients' money, and, in any case, may only be done within the provisions of Regulation 4(a). It may be thought, on the face of it, that Regulation 4(a) might give the necessary authority for a Member to 'top up' his client account, but this is not so; Regulation 4(a) is restricted to the nominal sum (if any) required to open the client account at a bank.

See also the Notes to Regulation 6.

(iii) A Member cannot treat himself as a client and in consequence cannot conduct his personal or office transactions through his client account.

'client'

(i) For the avoidance of doubt, it should be noted that each individual owner of a unit within a tenement property in Scotland is not to be regarded as the client for the purposes of these Regulations. The general body of proprietors of units in a tenement property shall normally be regarded as the client.

(ii) For the purposes of these Regulations it should also be noted that a client includes a person or body on whose account a Member holds or receives money as stakeholder.

'bank'

(i) Members may keep clients' money in a building society provided:

(a) the building society has trustee status, i.e. one which has been designated under Section 1 of the House Purchase and Housing Act, 1959; and

(b) any clients' money kept in such building society is arranged to be on immediate call.

(ii) It is for the Member, remembering that it is his clients' and not his own money involved, to decide whether or not to place funds with any particular bank or building society. The General Council regards it as essential that money in the client account should be immediately available for the use of clients and if, for any reason, clients' money which is in a deposit account is required, this also should be immediately available even at the sacrifice of interest.

(iii) Interest on deposits in estate agency transactions should be dealt with in accordance with the provisions of the Estate Agents Act 1979 and the Estate Agents (Accounts) Regulations 1981. Interest on other clients' money is, in the absence of agreement to the contrary, payable as a matter of law to the client. An expanded note on interest on clients' money appeared on page 603 of *Chartered Surveyor Weekly*, and page 320 of *Chartered Quantity Surveyor*, for May 1982.

Part II – Bank Accounts

Regulation 2

(1) Every Member who holds or receives clients' money shall without delay pay such money into a client account.

(2) Any Member who holds or receives clients' money must keep at least one client account or as many such accounts as he thinks fit.

Notes to Regulation 2

(i) Estate Agents Act 1979

Members who receive estate agency deposits must keep an appropriately identified separate clients' account for that purpose and ensure that the account and their records relating to such deposits conform with the requirements of the Act and the Estate Agents (Accounts) Regulations 1981 as well as with the Members' Accounts Regulations.

(ii) The qualifications in Regulation 8(1) will be noted.

(iii) Money held on a joint account by a Member with another Member or firm, for example on a joint stakeholder account, is not clients' money for the purpose of the Regulations, but it is money in respect of which a separate record should be kept.

(iv) It is open to a Member, within the requirements of the Regulations, to open any number of client accounts, whether for individual clients or for separately identifiable purposes e.g.:

J's Will Trusts Client Account
Client Management Account

The qualification in Regulation 8(1)(c) will be noted.

(v) It will be appreciated that in the event of bankruptcy or death of any Member, particularly a sole practitioner, difficulty might arise as to the identity of clients' money in a bank account not containing the word 'client'. The Institution has been advised by Counsel that monies held by Members on behalf of clients are impressed with a trust; and that failure to account for such monies would be likely to expose Members to an action for breach of trust.

Regulation 3

Where a Member holds or receives a cheque or bankers' draft which includes clients' money he shall pay such a cheque or draft into a client account.

Note to Regulation 3

The qualification in Regulation 5 will be noted.

Regulation 4

There may be paid into a client account:

(a) such money belonging to the Member as may be the minimum required by the bank for the purpose of opening or maintaining that account;

(b) money to replace any sum which for any reason may have been drawn from the account in contravention of these Regulations.

Notes to Regulation 4

4(a) See also Notes to Regulations 1 and 6.

Regulation 5

No money other than money which a Member is required or permitted to pay into a client account shall be paid into such an account, and it shall be the duty of a Member into whose client account any money has been paid in contravention of these Regulations to withdraw the same without delay.

Notes to Regulation 5

Money received which comprises both clients' money and money belonging to a Member should be paid into a client account and applied in accordance with Regulation 6(d).

Regulation 6

No money may be drawn from a client account except:

(a) money properly required for payment to or on behalf of a client;

(b) money properly required for or towards reimbursement of money expended by the Member on behalf of a client;

(c) money properly required for or toward payment of a Member's fees and/or disbursements

  (i) where there has been delivered to the client or his solicitor a bill for such fees and/or disbursements or other written intimation of the costs incurred and it has thereby or otherwise in writing been made clear to the client or his solicitor that money is being or will be applied towards or in satisfaction of such fees and/or disbursements; or

  (ii) where such drawings may have been expressly authorised in writing by a client beforehand;

(d) money (originally forming part of a larger sum paid into a client account by virtue of Regulation 3) to which the Member is beneficially entitled and such sum shall be withdrawn from a client account as soon as the net amount of the Member's entitlement can be quantified;

(e) money which is transferred to another client account;

(f) such money as may have been paid into an account under Regulation 4(a);

(g) such money which for any reason may have been paid into the account in contravention of Regulation 5.

Provided that money drawn on behalf of a client from a client account under sub-paragraphs (a), (b), (c), (d) or (e) of this paragraph shall not exceed the total of the money held at the time of such drawing on behalf of that client.

Notes to Regulation 6

(i) For a letter to be regarded as a written intimation of fees and/or disbursements a copy should be preserved in the same manner as, and with, other copy bills.

(ii) It should be noted that a client can authorise another person to receive bills for fees and/or disbursements on his behalf.

(iii) It should be noted that drawing money on behalf of a client from a client account when such drawing exceeds the total of the money held on behalf of that client is not permissible. In order to comply with this Regulation, in a situation where there are insufficient funds in an individual client's ledger to meet a particular payment, the procedures must be:

(a) either draw one cheque on the client account to the extent of the funds held on behalf of that particular client, and another cheque on the firm's accounts for the balance; or
(b) transfer sufficient firm's money to the client's ledger (with the appropriate transfer of funds to the client account), and then draw a cheque on the client account for the total sum required.

The underlying reason for this requirement lies in the rules which apply in the case of bankruptcy, since it is essential that clients' money can be identified as such.

(iv) Payments to or on behalf of a client should as far as possible always be made by means of a crossed cheque. In order to protect clients' funds against misappropriation where cheques are made payable to banks or building societies it is strongly recommended that the words 'account of . . .' or 'draft payable to . . .' be included after the payee.

(v) It will be appreciated that a Member should use caution in drawing against a cheque before it has been cleared as in the event of it being dishonoured a shortage of clients' funds would arise. The Member can avoid a breach of the Regulations by instructing his bank, if practicable, to charge all unpaid credits to his office or personal bank account.

Regulation 7

(1) No money drawn from a client account under Regulation 6 (b),(c),(d), (f) or (g) shall be drawn except by:

(a) a cheque drawn in favour of the Member, his firm or company; or
(b) a transfer to a bank account in the name of the Member, his firm or company, not being a client account.

(2) No money may be drawn from a client account otherwise than under the signature of at least one of the following:

(a) a Member;
(b) a partner or co-director of a Member;
(c) a Member of the Institution employed by a Member or his firm or company;
(d) an accountant who, but for his employment by a Member, or his

firm or company, would have been qualified to sign an accountant's report;

(e) in exceptional circumstances, a person not falling within (a), (b), (c) or (d) above expressly authorised by the Institution;

provided that any signatory referred to in (c), (d) or (e) above is the subject of dishonesty of employees cover in the Member's firm's or company's professional indemnity insurance policy.

(3) Notwithstanding anything contained in these Regulations the General Council may on the application of any Member or of its own motion authorise the withdrawal of money from a client account in circumstances which would otherwise not be permitted by these Regulations.

Notes to Regulation 7

(i) Regulation 7(2)(e) permits persons other than Members of the Institution, their partners/co-directors and qualified accountants employed by the firm to sign cheques drawn on client account provided the prior consent of the Institution is obtained. Such consent will normally be granted only in exceptional circumstances, since it is a cardinal principle of the Regulations that chartered surveyor principals shall be personally responsible for the handling by their firms of other people's money. Members must ensure that any person having this delegated authority is the subject of dishonesty of employees cover in the firm's professional indemnity insurance policy.

(ii) Regulation 7(3) gives the General Council a discretion to permit a Member or Members to withdraw client's money from the client account in circumstances where such withdrawal would not otherwise be permissable, e.g. where it can be shown to the satisfaction of the General Council that compliance with the proviso to Regulation 6 would involve a Member in more than normal difficulty or expense. However, the General Council does not anticipate granting such waivers, other than in the most exceptional cases.

Regulation 8

(1) Notwithstanding the provisions of these Regulations a Member shall not be under an obligation to pay into a client account money held or received by him:

(a) which is received by him in the form of cash and is without delay paid in cash in the ordinary course of business to a client or on his behalf to a third party; or

(b) which is received by him in the form of a cheque or draft which is endorsed over in the ordinary course of business to a client or on his behalf to a third party and is not passed by the Member through a bank account; or

(c) which he pays into a separate account opened in the name of a client or of some person designated in writing by a client.

(2) Notwithstanding the provisions of these Regulations, a Member shall not pay into a client account money held or received by him:

(a) which is received by him for or towards payment of a debt due to the Member from a client or in reimbursement of money expended by the Member on behalf of a client; or

(b) which is expressly paid to him for fees and/or disbursements incurred in respect of which a bill or other written intimation of the amount incurred has been delivered for payment, or as an expressly agreed fee or commission for business undertaken or to be undertaken; or

(c) which the client for his own convenience requests be withheld from such account, the request to be in writing from the client and acknowledged in writing by the Member.

Notes to Regulation 8

(i) 8(1)(a) and (b). The requirement of Regulation 10 concerning the need to record all movements of clients' money will be noted.

(ii) The type of account envisaged by Regulation 8(1)(c) is one over which a member has no control. For this purpose, a mandate signed by the client for the Member to draw on the account on behalf of the client does not constitute control unless the Member is the only person entitled to draw on the account.

(iii) It should be noted that payments received as envisaged in Regulation 8(2)(b) are not to be paid into a client account. The 'agreed fee or commission' must have been agreed in writing with the client. Money paid on account of fees and disbursements generally prior to the delivery of a bill or other written intimation or the written agreement of a fee should be paid into client account and transfer from client to office account can only take place when a bill or other written intimation has been delivered or a fee has been agreed in writing. For a letter to be regarded as a written intimation of fees and/or disbursements a copy should be preserved in the same manner as, and with, other copy bills.

Regulation 9

No sum shall be transferred from the account of one client to that of another except in circumstances in which it would have been permissible under these Regulations to have withdrawn from a client account the sum transferred from the first client and to have paid into a client account the sum transferred to the second client.

Part III – Books and Records

Regulation 10

(1) Every Member shall at all times keep properly written up such accounts as may be necessary:

(a) to show his dealings with:

(i) all clients' money received, held or paid by him; and

(ii) any other money dealt with by him through a client account;

(b) to show separately in respect of each client all clients' money which is received, held or paid by him on account of such client.

(2) All dealings referred to in sub-paragraph (a) of paragraph (1) of this Regulation shall be recorded as appropriate: either

(a) in a clients' cashbook, or in a client column of a cash book; or

(b) in a record of sums transferred from the ledger account of one client to that of another;

and in addition, in a clients' ledger, or in a clients' column of a ledger.

(3) Every Member shall not less than once in every succeeding period of three months cause the balance of his clients' cash book (or clients' column of his cash book) to be agreed with his client account bank statements and with his clients' ledger balances and shall keep in the cash book or other appropriate place a reconciliation statement showing this agreement.

(4) In this Regulation the expressions 'accounts', 'books', and 'ledgers' shall be deemed to include looseleaf books and such cards or other permanent documents or records as are necessary for the operation of any system of book-keeping, mechanical or otherwise, including computer-operated systems.

(5) A register of all clients' accounts must be maintained.

Notes to Regulation 10

(i) It is considered to be good accountancy practice that the balance of clients' cash book should be agreed with the client account bank statements and with clients' ledger balances monthly.

(ii) A cheque made payable to a Member or his firm and endorsed over to a third party constitutes a dealing with clients' money and should be recorded as a payment in and payment out.

Part IV – Monitoring and Enforcement

Regulation 11

(1) Once in each period of twelve months every Member who is a sole principal of a practice or a partner in a firm or a director of a company or is held out to the public as a partner in a firm or a director of a company carrying on practice as surveyors in the United Kingdom, the Isle of Man or the Channel Islands shall deliver to the Institution:

(a) a certificate signed by the Member that during the accounting period covered thereby he did or did not receive or hold clients' money; and, if he did,

(b) an accountant's report in the form set out in the Schedule to the Accountant's Report Regulations relating to the accounting period.

(2) The Certificate and, where relevant, the report under paragraph (1) of this Regulation shall be delivered to the Institution not more than six months after the end of the accounting period to which it relates.

(3) The accounting period:

(a) shall begin at the expiry of the last preceding accounting period for which a certificate under paragraph (1) of this Regulation has been delivered;

(b) shall cover not less than twelve months; and

(c) shall where possible correspond to a period or consecutive periods for which the accounts of the Member are ordinarily made up.

Provided that in respect of a Member who commences or re-commences to practise as a surveyor either as a sole principal or as a partner in a firm or as a director of a company, or in respect of the first certificate and, where appropriate, accountant's report delivered after these Regulations shall have come into force, the accounting period may be for a period of less than twelve months.

(4) In the case of a Member who has two or more places of business:

(a) separate accounting periods covered by separate certificates and accountant's reports may be adopted in respect of each place of business; and

(b) the accountant's report or reports delivered by him in each year shall cover all clients' money held or received by him.

(5) The requirements of paragraph (1) of this Regulation shall in the case of a firm or company carry on practice as surveyors be deemed to be satisfied in respect of all Members who are named in one certificate or accountant's report as being partners or directors of the firm or company concerned and who shall have signed the said certificate.

(6) Notwithstanding the generality of sub-paragraph (1) of this Regulation, a Member who submits a certificate to the effect that he did not receive or hold clients' money during the accounting period covered thereby may be exempted from the requirement to deliver annual certificates for such period as the General Council may determine unless:

(a) before the expiry of such period of exemption he receives or holds clients' money, when he shall forthwith inform the Institution accordingly; or

(b) the General Council decides to terminate an exemption previously granted, which it shall have the power to do at any time during the period of the exemption.

Regulation 12

(1) In order to ascertain whether these Regulations have been complied with, the Professional Conduct Committee of the Institution acting either:

(a) on its own motion; or

(b) on being satisfied on a written complaint being lodged with it by a third party that there is prima facie evidence that a ground of complaint exists,

may require a Member to produce at a time and place to be notified to the Member by or on behalf of the Secretary-General his books of account, bank pass books, looseleaf bank statements, statements of account, vouchers including petty cash vouchers and any other necessary documents for the inspection of any person appointed by the Institution and to supply to such person any necessary information or explanation and such person shall be directed to prepare for the information of the Institution a report of such inspection. Such report may be used as a basis for disciplinary proceedings under Bye-Law 24.

In cases of urgency and/or in circumstances when it is not practicable to delay any action until the next formal meeting of the Professional Conduct Committee, the production of documents and the supply of information pursuant to this Regulation can be required upon the direction of three or more members of the Committee, including at least one of its Officers, acting on its behalf.

(2) Either the Professional Conduct Committee or the Disciplinary Board or the Appeals Board, as the case may be, shall have power to make such order as it shall consider just for payment by the Member of a sum of money in or towards payment of the costs incurred by the Institution in connection with any inspection made by any person appointed under this Regulation.

Regulation 13

A requirement to be made of a Member under these Regulations shall be made by or on behalf of the Secretary-General and shall be sent by recorded post to the Member at his last-known address.

Regulation 14

The General Council shall have the power to waive in any particular case any of the provisions of these Regulations.

Appendix VII

RICS CLIENTS' MONEY PROTECTION SCHEME

The clients' money protection scheme aims to provide for any member of the public, which includes any person, firm, trust, body corporate or other organisation, to be reimbursed his direct pecuniary loss arising from the fraud or dishonesty of, or misappropriation by, any corporate member of the Royal Institution of Chartered Surveyors and/or any principal or employee of any qualified firm carrying on practice as surveyors in the British Isles subject to the following notes.

For the purposes of the scheme, 'direct pecuniary loss' is defined as being the loss of principal monies only entrusted to a Chartered Surveyor or qualified firm and excludes all amounts due, or which might be due, in respect of interest thereon or loss of profits or other loss in any way arising therefrom, and also excludes monies deposited with a Chartered Surveyor or qualified firm for the purpose of lodgement with any building society or other financial institution as an investment transaction.

For the purposes of the scheme, a 'qualified firm' is one in which at least one principal (partner or director) or whose sole principal is a member of the Royal Institution of Chartered Surveyors.

The current financial limits of cover are £30,000 per claimant subject to an overall limit on payments in any year of £2,000,000.

The scheme is designed to afford some protection to members of the public by making funds available for their reimbursement when all other avenues are exhausted. Chartered Surveyors and qualified firms can themselves receive no financial benefit or protection under the scheme. In the event of a loss the primary obligation is on the defaulting member or qualified firm to make full restitution. It is only in cases where the member or firm concerned is unable to do so that the scheme can take effect.

It is a precondition of the admission of any claim under the scheme that the name of the alleged defaulter shall be reported to the police authorities on behalf of the claimant. All claims should then be submitted in writing to the Secretary-General, at Surveyors Bookshop, 12 Great George Street, Parliament Square, London SW1P 3AD.

MARCH 1989
MG/MW

Appendix VIII

EXTRACT FROM RICS GUIDANCE RELATING TO CONNECTED BUSINESSES

The Royal Institution of Chartered Surveyors Operation of Regulations 6(a)

1. A holding of less than 25% of voting shares, provided the Member was not otherwise involved in the management or control of the company, would probably be permitted. A holding of between 25% and 50% could concern the Institution depending on the allocation of the remaining equity; and a holding of 51% or more would be likely to be prohibited.
2. A Member holding a directorship or partnership in a contracting business would certainly be deemed to be involved in the management or control of the business since he could be held liable for it. Any other management position in the business would be looked at on its facts. The relevant considerations would include:
   - whether the Member was 'held out' as being a partner or director of the business, such that he might be held liable in law for its conduct;
   - whether the Member stood to gain directly from the financial success of the business;
   - whether the Member had access to, or was in a position to influence, tenders prepared by the business;
   - whether the Member had any control over what work was or was not accepted by the business.

3. Limits of Restriction

Many of the enquiries received by Members concern proposed arrangements which would not be prohibited. For example, a Member is not precluded by Regulation 6(a) from any of the following:

(a) He may take part in the management or control of a contracting business provided that he is not at the same time carrying on practice as a surveyor.

(b) He may take part in the management or control of a contracting business which also provides some services normally associated with surveying through 'design and build' contracts, but with the same proviso as in (a).

(c) He may engage in building development projects as a principal, for example, through his own property development or investment company even if he is at the same time in practice as a surveyor.

(d) He may act as consultant to a contracting business while at the same time in, or as part of, practice as a surveyor. Here the relationship with the contracting company will be akin to that with a client.

(e) He may hold part of the equity in a contracting business, while at the same time in practice as a surveyor, provided that his holding does not give him effective control over the contracting business. Each case on which the Institution is consulted would be assessed for compliance on its own facts (see paragraph 1 above).

(f) A Member in practice as a surveyor who does not perform the function of quantity surveying may take part in the management or control of a business engaged 'partly' (as opposed to 'substantially'), in contracting: for example, a business which is primarily engaged in a non-contracting function but which owns a contracting subsidiary; or one of which contracting forms a small proportion of the company's turnover.

(g) A Member in practice as a surveyor who does not perform the function of quantity surveying may directly employ workmen for the purpose of carrying out maintenance and repairs to clients' properties under management by his practice, and (provided that this forms a non-substantial part of the business) to other properties.

1988

Appendix IX

SALARIED PARTNERSHIP AGREEMENT

This suggested sample agreement must not be used without seeking professional legal, tax and accountancy advice. It is intended only to serve as a guide and is reproduced for illustrative purposes only. It is not intended to be a model.

T H I S AGREEMENT FOR SALARIED PARTNERSHIP is made the day of 199 B E T W E E N the Partners in the Firm of of of the one part and ______________ of ('the Salaried Partner') of the other part

W H E R E A S the Salaried Partner is a member of the staff of the Firm and the terms of the Salaried Partner's Contract of Employment are set out in the Schedule hereto

AND W H E R E A S the Firm has invited the Salaried Partner to be associated with the Firm upon the terms and conditions hereinafter contained

N O W IT IS HEREBY AGREED as follows:

1. IN this Agreement the following expressions shall where the context permits have the following meanings:

   (1) 'the Partners' means the persons (other than the Salaried Partners) who shall for the time being be partners in the Firm and 'Partner' has a corresponding meaning

   (2) 'the Salaried Partners' means the Salaried Partner and the persons who shall for the time being be Salaried Partners in the Firm being persons who are not entitled to share in the profits nor liable to suffer any losses of the Firm and 'Salaried Partner' has a corresponding meaning

   (3) 'the Firm' means the partnership which subsists between the Partners carrying on the practice of surveyors at under the style or firm name of or under such other name or names as may from time to time be agreed by the Partners

2. ON and from the day of 199 the Salaried

Partner shall be associated with the Firm as a Salaried Partner for the period of One (1) year and thereafter until terminated by not less than six (6) months' prior written notice expiring on the last day of any calendar month given by the Salaried Partner or the Firm to the other

3. THE existing Partnership Agreement between the Partners shall inter se remain in full force and effect save only as modified by this Agreement but any variation of the provisions of such Partnership Agreement or any new partnership agreement subsequently entered into by the Partners shall not in any way prejudicially affect the rights of the Salaried Partner hereunder

4. THE Partners shall pay and discharge all liabilities of the Firm (other than liabilities incurred by the Salaried Partner on behalf of the Firm while the Salaried Partner is acting contrary to instructions or in breach of the terms of this Agreement) and shall keep the Salaried Partner and the Salaried Partner's personal representatives estate and effects fully and effectually indemnified against all such liabilities and against all claims proceedings costs damages and expenses in respect thereof and for the avoidance of doubt IT IS HEREBY AGREED that:

(1) no Partner shall be accountable under this clause 4 in respect of any liability arising after such Partner shall cease to be a Partner

(2) notwithstanding Section 18 of the Partnership Act 1890 or any statutory replacement or modification thereof for the time being in force no change in the constitution of the Firm shall impair or discharge the liability of the Partners (including such other person or persons as shall be admitted after the date hereof as a Partner or Partners in the Firm) under the agreement and indemnity contained in this clause 4 except as provided in sub-clause (1) hereof

5. THE Salaried Partner shall not during the Salaried Partner's association as a Salaried Partner with the Firm without the consent of the Firm do any of the things following:

(1) give any security or undertaking for the payment of money on account of the Firm other than such undertakings as are normally given in the ordinary course of business by a surveyor

(2) release or compound any debt owing to or claimed by the Firm for an amount exceeding One hundred pounds (£100)

(3) sign draw endorse or negotiate any bill promissory note bond or other security or become surety for any person or do or knowingly permit to be done anything by which the capital of the Firm may be charged taken in execution or put in jeopardy

(4) lend any money belonging to or give any unusual credit on behalf of the Firm

6. (1) THE Salaried Partner shall bring into account all directors' fees

and other remuneration and emoluments payable to the Salaried Partner by any company or person except in respect of appointments which shall have been disclosed to and agreed by the Firm

(2) Any legacy or gift to the Salaried Partner being an express or indirect return in lieu of professional charges for services rendered by the Firm shall not belong to the Salaried Partner and shall be accounted for to the Firm

7. (1) THE Firm shall include the Salaried Partner's name on the Firm's headed notepaper and the Partners ACKNOWLEDGE AND AGREE that the Salaried Partner shall be entitled to act as a Partner for the purpose of carrying on the usual business of the Firm

(2) The Salaried Partner shall have no rights or liabilities in respect of the management of the Firm but it is the policy of the Partners to seek and welcome the advice views and support of Salaried Partners in the spheres of management in which Salaried Partners are able to make contributions to the well-being of the practice Further the Salaried Partner may sign cheques upon the Firm as agreed by the Partners within such limits as the Partners may from time to time set

(3) Nothing herein contained shall be construed or have effect as entitling the Salaried Partner to share in any profits of the Firm and the Partners ACKNOWLEDGE AND AGREE that all losses of the Firm shall be borne by the Partners

8. THE association of the Salaried Partner as a Salaried Partner in the Firm (in contrast with employment by the Firm as a surveyor) may be determined forthwith by notice in writing given by the Firm to the Salaried Partner:

(1) if the Salaried Partner shall:

(a) commit any serious breach of stipulations herein contained or

(b) commit any act of bankruptcy or

(c) commit any criminal offence other than an offence under The Road Traffic Acts or

(d) in the opinion of the Firm behave in a manner inconsistent with that which behoves a person held out as a Partner of the Firm or

(e) without due cause fail to account for and pay over or refund any moneys for which the Salaried Partner is accountable to the Firm within ten (10) days after being required so to do in writing by any Partner of the Firm or

(f) act in any other respect contrary to the good faith which ought reasonably to be observed between Partners or

(g) cease to be employed by the Firm

(2) on the sale transfer dissolution or partition of the practice of the Firm or any part thereof

whereupon the rights and obligations created under this Agreement as to the association of the Salaried Partner as a Salaried Partner in the Firm shall thereupon be determined

9. IF the Partners shall so require then the Salaried Partner (or the Salaried Partner's personal representatives) shall join with the Partners in giving H.M. Inspector of Taxes (or the Special Commissioners of Taxes) a notice under Section 154(2) of the Income and Corporation Taxes Act 1970 or any statutory replacement or modification thereof for the time being in force and the Salaried Partner (or the personal representatives of the Salaried Partner) shall be indemnified by the Partners against any tax which may be payable by the Salaried Partner or the Salaried Partner's personal representatives as the result of giving such notice in excess of the tax which would have been payable if no such notice had been given

10. ANY dispute which may arise concerning anything contained in or provided by this Agreement shall be referred to a sole arbitrator pursuant to the provisions in that behalf of the Arbitration Acts 1950 and 1979 or any statutory modification or re-enactment thereof for the time being in force such arbitrator to be nominated by mutual agreement and in default of agreement by the President or one of the Vice Presidents for the time being of The Royal Institution of Chartered Surveyors

11. IT is the intention of the Partners of the Firm to pay commission to the Salaried Partner during the Salaried Partner's association as Salaried Partner with the Firm Commission shall be of such amount as the Partners of the Firm shall from time to time decide but in default of any decision to the contrary shall be calculated and paid as follows PROVIDED THAT this clause shall not be capable of being construed as creating any legal obligation on the Partners of the Firm to pay commission:

(1) commission shall be calculated and equivalent to per cent (%) of the net profit of the Firm the net profit of any accounting period of the Firm falling only partly within the period during which the Salaried Partner shall be in association as Salaried Partner with the Firm being apportioned on a daily basis to ascertain the proportion of that net profit attributable to the period of the association

(2) commission shall be payable within twenty-one (21) days after the accountant to the Firm shall have certified the amount of the net profit for the relevant accounting period of the Firm

(3) for the purposes of this clause "net profit" shall mean the profit of the Firm before deduction of income tax or any other taxes

payable in respect of the profits of the Firm but after taking into account all usual charges expenses salaries (including any over-riding salaries to individual Partners in the Firm and the salary of the Salaried Partner) interest on the capital of Partners of the Firm services commissions (but excluding the commission payable to the Salaried Partner) all rents notional rents rates and other outgoings and such sum or sums as the accountant to the Firm shall consider reasonable and proper and to be in accordance with the accountancy practice of the Firm including work in progress bad and doubtful debts and depreciation

(4) the written certificate of the accountant to the Firm as to the amount of the net profit shall be final and binding on the parties hereto and the Salaried Partner shall not be entitled to see or receive copies of the Firm's accounts

12. IF the context so admits the masculine gender includes the feminine gender

AS WITNESS the hands of the parties hereto the day and year first before written

THE SCHEDULE

Terms of Contract of Employment

1. DATE: 199

2. EMPLOYERS: Surveyors of ('the Firm').

3. SALARIED PARTNER: Name:
Address:
Telephone No. (if any):
Date of Birth:

4. DATE OF COMMENCEMENT OF CONTINUOUS PERIOD OF EMPLOYMENT: 199

5. TITLE OF EMPLOYMENT: Surveyor

6. REMUNERATION: Not less than £ per annum payable by monthly instalments in arrear on the last day of each month (or earlier according to the Firm's accounting practice as shall from time to time be followed) which remuneration shall be deemed to accrue from day to day payment being made by cheque or bank credit transfer unless otherwise agreed. It is the policy of the Firm to review salaries periodically as circumstances shall dictate and in any event not less than once in every 12 months.

7. HOURS OF WORK: Monday to Friday: a.m. to p.m. with a break for lunch of hours to be taken between p.m. and p.m. except as may be reasonably necessary or desirable in the course of the Salaried Partner's duties.
The Salaried Partner shall devote the Salaried Partner's whole time and attention and diligently and to the best of the Salaried Partner's skill and ability perform the duties of the employment during the hours of work.
The Salaried Partner shall attend at the offices of the Firm or elsewhere during and outside these times as may be reasonably required or necessary for the proper discharge of the duties of the employment or in the event of the absence of others or due to the need to meet the urgency or importance of particular work in hand PROVIDED THAT this clause shall not create any obligation on the Salaried Partner to change the Salaried Partner's normal place of employment on a permanent basis.

8. HOLIDAY ENTITLEMENT: Bank and other public holidays and, in addition (on dates as shall first be notified and agreed after discussion with the partner concerned) in each year commencing 1st January, fully paid basic holidays of 20 working days of which not more than 15 working days shall normally be taken in the months of May to September (inclusive).
Holiday entitlement due in respect of employment during any part of a year commencing 1st January shall be deemed to have accrued and be calculated pro rata.
Entitlement shall be increased to 25 working days in each year with effect from the year commencing 1st January after 5 years of employment with the Firm.

9. SICKNESS AND INJURY PAY: The Statutory Sick Pay Scheme ('SSPS') applies to this employment. The Firm will pay to the Salaried Partner during absence through sickness or injury (after taking into account sums to which the Salaried Partner is entitled under the SSPS):
   (1) for the first 130 working days pay at the same rate as the Salaried Partner's remuneration,
   (2) thereafter at the Firm's discretion.

For any period of incapacity for work of between 4 and 7 calendar days a self-certification form must be signed and handed to the office manager. The Firm may require such forms to be signed for lesser periods at its discretion. For any longer period of incapacity for work a doctor's statement must be obtained and handed to the office manager. Absences from work should be notified on the first day of absence.

For the purposes of the SSPS the 'qualifying days' are Monday to Friday inclusive.

10. RIGHTS TO NOTICE ON TERMINATION OF EMPLOYMENT: Whether the Contract be for a fixed term (in which case it shall not be terminable before the expiration of the fixed term except as hereinafter provided) or otherwise not less than 6 months' prior written notice expiring on the last day of any calendar month shall be given by the Salaried Partner or the Firm to the other of the termination of the employment and the employment shall continue until so terminated.

PROVIDED THAT the Firm may terminate the Contract of Employment forthwith (in which event the Salaried Partner shall not be entitled to any further payment except sums as shall have accrued due) if the Salaried Partner shall:

(1) fail to observe the Disciplinary Rules set out in paragraphs 11(1)(a) and (b) of this Statement,

(2) be in breach of any Disciplinary Rule or fail to perform efficiently the duties of the employment subsequent to the receipt by the Salaried Partner of not less than 1 verbal warning followed by 1 written warning in relation to earlier breaches or failure as aforesaid,

(3) be prevented from the proper performance of the duties of the employment by the Salaried Partner's illness, injury or other cause (otherwise than arising through holiday entitlement or time off or leave to which the Salaried Partner is by law entitled):

(a) during the first 8 years of service for an aggregate of 130 working days in any continuous period of 12 months,

(b) after 8 years of service for an aggregate of 260 working days in any continuous period of 13 months.

No notice to terminate the Contract of Employment shall be given under this sub-clause (3) until the expiry of 1 month (in the case of continuous service not exceeding 8 years) and 2 months (in the case of continuous service of 8 years or more) after the giving of notification in writing to the Salaried Partner that the Contract may be terminated due to absence from work. Such notification shall not be given until there shall have been absence from work in any relevant period and an absence from work of a further 20 working days or less would give rise to the right under this sub-clause to terminate the Contract or such right shall have already arisen, or

(4) be guilty of dishonesty whether or not during the course of employment.

The normal retiring age of salaried partners in the Firm is 65 years for men and 60 years for women and the Firm's Pension Scheme has been devised with this factor in mind.

11. DISCIPLINARY RULES:

(1) (a) Not to disclose to any person or make public during or at any time after employment:

(i) the affairs of the Firm's clients nor indeed that the Firm acts for any particular client, except as shall be proper and necessary in the course of employment,

(ii) the methods of work or costing, details of organisation, precedent forms of the Firm or the practice, business dealings or affairs generally of the Firm, except as shall be proper and necessary in the course of employment.

(b) To take all steps in the care and control of client matters in respect of which the Salaried Partner has responsibility which shall be necessary to comply with professional standards, the breach of which might result in disciplinary proceedings against the Firm or any partner or employee (including the Salaried Partner) of the Firm.

(c) To accept and apply procedures and changes in such procedures and in equipment and machinery as are reasonable.

(d) Neither alone nor in conjunction with any other person, firm or company to engage directly or indirectly in any profession, trade or business other than that of the Firm except with the prior consent of the Firm and except also any matter or business which is performed without remuneration and is either carried out for relatives of the Salaried Partner or is not a matter or business normally dealt with by surveyors

(e) To keep or cause to be kept all such records as are usual and proper to be kept by the Salaried Partner in the performance of the Salaried Partner's duties including a record of all moneys paid and received in respect of any business or matter to which the Salaried Partner shall be entrusted and of all business and transactions undertaken or transacted by the Firm under the Salaried Partner's management or otherwise as may be required by the Firm.

(f) Without the consent of the Firm neither directly nor otherwise to be party to or involved in dealings in real, leasehold or other immovable property except by way of investment of a permanent nature or as a residence for the Salaried Partner and the Salaried Partner's family.

(2) On the making of a disciplinary decision the Salaried Partner shall be entitled:

(a) to an opportunity to state the Salaried Partner's case at the time it is under consideration,
(b) to appeal against the decision when it has been made to the senior partner of the Firm (or 2 other partners if the senior partner shall be directly involved in the case or not readily available),
(c) to be accompanied in either of the previous 2 situations by a colleague or other representative of the Salaried Partner's choice.

12. GRIEVANCE PROCEDURE: Any grievance relating to the employment shall be raised with the senior partner of the Firm or any other partner of the Firm and shall be expressed in writing if so required.

13. SOCIAL SECURITY PENSIONS AND PENSION SCHEMES: The Salaried Partner shall be a contributor to the Social Security Pension Scheme.
The Salaried Partner a member of the Firm's Pension Scheme, making such contributions, being entitled to such benefits and subject to such conditions as are set out in the Rules of the Scheme from time to time in operation, copies of which Rules being obtainable by application to the office manager.

14. For a period of 3 years after the termination of the employment the Salaried Partner will not without the prior written consent of the Firm directly or indirectly solicit, canvass or entice away any firm or person who now is, or may during the subsistence of this Contract become or be a client of the Firm other than relatives and personal friends of the Salaried Partner introduced to the Firm by the Salaried Partner as shall be agreed between the Salaried Partner and the Firm.

15. For a period of 3 years after the termination of the employment the Salaried Partner will not without the prior written consent of the Firm directly or indirectly enter into or be engaged in the service of, or on the Salaried Partner's own account act as a principal of or as clerk, agent or assistant to or solicit or endeavour to obtain business for any Surveyors' practice or firm of surveyors, from a place of business situate within a radius of 3 miles of

16. The Salaried Partner shall observe the provisions of clause 15 as if they were repeated in this clause but with the substitution of for the reference to

17. OTHER PROVISIONS:
(1) The Firm may terminate the Contract of Employment without

notice (in which event the Salaried Partner shall not be entitled to any further payment except sums as shall have accrued due) if the Salaried Partner shall be adjudicated bankrupt or behave, whether or not during the course of employment, in a manner inconsistent with that which behoves a Surveyor with the Firm.
(2) The terms of the Contract of Employment set out in the Schedule to this Agreement are the entire terms of the employment of the Salaried Partner by the Firm and being by mutual consent are in substitution for any previous agreement of employment.

SIGNED by a Partner duly
authorised on behalf of the
Partners in the presence of: }

Witness'
signature ....
Name (capital
letters) ....
Address ....
....
....
....
Occupation ....

SIGNED by the Salaried
Partner in the presence of: }

Witness'
signature ....
Name (capital
letters) ....
Address ....
....
....
....
Occupation ....

Appendix X

EQUITY PARTNERSHIP AGREEMENT

This sample agreement must not be used without seeking professional legal, tax and accountancy advice. It is intended to serve as a guide and is reproduced for illustrative purposes only. It is not intended to be a model.

THIS AGREEMENT is made the

day of 1989 BETWEEN SUSAN OCTAVIA CAREFREE Bachelor of Science, Associate of The Royal Institution of Chartered Surveyors of The Manor House Anytown ('Miss Carefree') of the first part MANFRED ERIC JOLLY Master of Arts, Fellow of the Royal Institution of Chartered Surveyors of The Grange Anytown ('Mr. Jolly') of the second part and MELANIE REBECCA HAPPY Associate of the Royal Institution of Chartered Surveyors of The Manse Anytown ('Mrs. Happy') of the third part and IAN MARK HOPEFUL Associate of The Royal Institution of Chartered Surveyors of 22 Green Street Anytown ('Mr. Hopeful') of the Fourth part

WHEREAS
(1) Miss Carefree Mr. Jolly and Mrs. Happy ('the Continuing Partners') have for some years past carried on in partnership the business of Chartered Surveyors ('the business') from premises at 1 High Street Anytown ('the Partnership Premises')
(2) The Continuing Partners have agreed to take Mr. Hopeful into partnership in the business with effect from the Sixth day of April 1990 on the terms hereinafter set out

NOW IT IS HEREBY AGREED as follows:

1. THE parties hereto ('the Partners') shall as from the Sixth day of April 1990 carry on the partnership on the terms of this Agreement

2. THE business name shall be 'Carefree Jolly & Happy' and the Partners shall comply at all times with the requirements of the Business Names Act 1985 so far as the same relate to the partnership

3. SUBJECT to the provisions for dissolution hereinafter contained the partnership shall continue for the joint lives of the Partners

4. THE business of the partnership shall be carried on from the Partnership Premises and at such other place or places as the Partners may from time to time agree

5. (1) THE initial capital of the partnership shall be the total sum shown as capital in the books of the previous partnership as at the Fifth day of April 1989 which sum shall belong to the Partners in the proportions in which it has been contributed by the Partners respectively
(2) The capital for the time being of the partnership shall belong to the Partners in the proportions in which it has been contributed by the Partners
(3) Should the Partners decide at any time to increase the capital of the partnership the amount of any increase shall be contributed unless the Partners shall have otherwise agreed in the same proportions as the Partners are then entitled to share in the profits and losses of the partnership
(4) Each of the Partners shall be entitled to interest at the rate of Twelve per cent (12%) per annum on the amount for the time being of such Partner's share of the partnership capital such interest to be credited each year before the profits are divided

6. (1) Miss Carefree shall be paid a salary of THIRTY THOUSAND pounds (£30,000.00) per annum and Mr. Jolly shall be paid a salary of TWENTY THOUSAND pounds (£20,000.00) per annum such salaries to be paid monthly in arrears PROVIDED THAT if in any year the profits of the partnership before payment of the said salaries shall be less than FIFTY THOUSAND pounds (£50,000.00) the said salaries shall be reduced proportionately
(2) Subject to the payment to Miss Carefree and Mr. Jolly of the salaries referred to in paragraph (1) of this clause the profits of the partnership including profits of a capital nature shall be divided between the Partners in the following proportions and the Partners shall bear all losses including losses of capital in the same proportions:
Miss Carefree THIRTY per cent
Mr. Jolly THIRTY per cent
Mrs. Happy TWENTY FIVE per cent
Mr. Hopeful FIFTEEN per cent

7. PROPER books of account shall be kept properly posted up and such books shall be available at all times for inspection by each of the Partners

8. ON the Fifth day of April in each year an account shall be taken of all assets and liabilities of the partnership and a balance sheet and profit and

loss account showing what is due to each Partner in respect of capital and share of profits and salary shall be prepared and shall be signed by each Partner who shall be bound thereby unless some manifest error shall be found therein within three (3) months in which case such error shall be rectified

9. THE accountants to the partnership shall be Messrs. Ledger Total and Balance or such other firm of chartered accountants as the Partners may from time to time agree

10. THE Partners shall be entitled to draw out of the partnership bank account on account of the Partners' respective salaries or shares of the profits such sums as shall be agreed between the Partners but if when the next following yearly account is taken it appears that any Partner has drawn any sum in excess of such Partner's salary or share of the profits such Partner shall forthwith repay such excess

11. (a) THE bankers of the partnership shall be Midland Bank PLC at their branch at Anytown or such bankers as the Partners may from time to time agree
(1) All partnership moneys not required for current expenses shall immediately upon receipt be paid into the said banking account and all cheques on such account shall be signed by Two of the Partners

12. EACH Partner shall devote such of such Partner's time and attention to the business of the partnership as shall be reasonably necessary for the proper conduct of the partnership business and shall use such Partner's best endeavours to promote the success of the partnership business

13. NONE of the Partners shall without the consent of the others:
(1) engage directly or indirectly or be concerned or interested in any business other than that of the partnership
(2) engage directly or indirectly or be concerned or interested in any business which competes directly with that of the partnership
(3) enter into any bond or become bail or surety for any person or knowingly cause or suffer to be done anything whereby the partnership property may be taken in execution or otherwise endangered
(4) assign mortgage or charge such Partner's share in the assets or profits of the partnership
(5) compromise or compound or (except upon payment in full) release or discharge any debt due to the partnership

14. IF any Partner:

(1) shall be adjudicated bankrupt or shall suffer such Partner's share in the partnership to be charged for such Partner's debt under the Partnership Act 1890 or

(2) shall become a patient within the meaning of the Mental Health Act 1959 or

(3) shall be suspended or expelled from Membership of The Royal Institution of Chartered Surveyors or

(4) shall commit any grave breach or persistent breaches of this Agreement

such Partner may be expelled from the partnership upon service on such Partner by the other Partners of a notice in writing dissolving the partnership

15. ON the death of any Partner or if and whenever any of the Partners is expelled from the partnership in accordance with the provisions of clause 14 hereof then if the continuing Partners shall not all elect by notice in writing to acquire the share of the deceased or expelled Partner ('the outgoing Partner') within thirty (30) days of the date of the outgoing Partner's death or expulsion the partnership shall be wound up in accordance with the provisions of the Partnership Act 1890 PROVIDED THAT if the continuing Partners do so elect as aforesaid then the share of the outgoing Partner in the capital and assets of the partnership shall vest in the continuing Partners in proportion to their respective shares in the profits of the partnership SUBJECT TO the payment by the continuing Partners to the outgoing Partner (or to the outgoing Partner's personal representatives in the case of a deceased Partner) of a sum or sums calculated in accordance with the following provisions:

(1) the continuing Partners shall cause a profit and loss account and balance sheet of the partnership business to the date of death or expulsion ('the succession date') to be drawn by the accountants to the partnership

(2) the outgoing Partner shall be paid a capital sum equal to the amount standing to the credit of the outgoing Partner's capital account as shown in the said account

(3) the outgoing Partner shall be paid any undrawn balance of the outgoing Partner's share of the net profits of the partnership for the period covered by the said account

(4) the outgoing Partner shall be paid any unpaid interest which has accrued on the outgoing Partner's capital to the succession date

(5) all necessary valuations shall be made by two (2) firms of independent valuers appointed by the outgoing Partner (or the outgoing Partner's personal representatives in the case of a deceased Partner) and the continuing Partners respectively

(6) no allowance shall be made for the goodwill of the business
(7) the continuing Partners shall indemnify the outgoing Partner (or the outgoing Partner's personal representatives in the case of a deceased Partner) from and against all existing and future debts liabilities and engagements of the partnership except any debt or liability in respect of income tax attributable to the outgoing Partner's share of the profits of the partnership
(8) the liability of the continuing Partners to make the foregoing payments shall be joint and several but as between the continuing Partners shall be borne in the proportions in which the share of the outgoing Partner is vested in the continuing Partners in accordance with the provisions of this clause
(9) all necessary and proper instruments shall be executed for vesting the share of the outgoing Partner in the continuing Partners and each of the Partners HEREBY IRREVOCABLY APPOINTS the other Partners and each of them such Partner's attorney for that purpose

16. THE capital sum equal to the amount standing to the credit of the outgoing Partner's capital account referred to in clause 15 hereof may at the option of the continuing Partners be paid by four (4) equal instalments the first to be paid within six (6) months of the succession date and the remaining instalments at intervals of every six (6) months thereafter with interest to the date of payment in each case on the balance of the said capital sum then outstanding at the rate which exceeds by Two per cent (2%) the Midland Bank PLC base rate for the time being

17. (a) ANY Partner may serve on the remaining Partners a notice that such Partner wishes to retire from the partnership at the expiration of three (3) months from the date of such notice and the remaining Partners may then elect by notice in writing addressed to the retiring Partner at any time prior to the expiration of such period of three (3) months that the retiring Partner's share shall vest in the remaining Partners at the expiration of such period in accordance with the provisions of clause 15 hereof and as if the retiring Partner were an outgoing Partner as defined in that clause
(1) If the remaining Partners shall not so elect that the retiring Partner's share shall vest in the remaining Partners the partnership shall be wound up at the expiration of the said period of three (3) months in accordance with the provisions of the Partnership Act 1890

18. ANY two (2) Partners or any one (1) of the Partners if there shall be only two (2) may determine the partnership by three (3) months' notice in

writing to the other Partners or Partner and at the expiration of such period the partnership shall be wound up in accordance with the provisions of the Partnership Act 1890

19. ON the retirement of any Partner or if any Partner shall be expelled from the partnership then unless the partnership shall thereupon be wound up in accordance with the provisions of the Partnership Act 1890 the outgoing Partner shall covenant not to carry on any business which is similar to or likely to compete with the partnership business whether alone or in partnership with or as servant or agent for any other person for a period of Five (5) years from the succession date and within a radius of Thirty (30) miles of the Partnership Premises

20. AT the request of the continuing Partners the outgoing Partner shall join with the continuing Partners and with any person or persons who may enter into partnership with the continuing Partners on the date of dissolution in making any election available under sub-section (2) of Section 154 of the Income and Corporation Taxes Act 1970 The continuing Partners shall keep the outgoing Partner and the outgoing Partner's estate and effects indemnified against the amount of all income tax (whether at the basic or any higher rate) suffered by the outgoing Partner which the outgoing Partner would not have suffered if the outgoing Partner had not joined in making the election

21. SO long as Miss Carefree shall be a Partner under the terms of this Agreement and shall be the owner of the Partnership Premises Miss Carefree shall permit the partnership to have the use and occupation of the Partnership Premises for the purposes of its business free of rent but upon the partnership discharging all rates and other outgoings in respect of the Partnership Premises

22. (a) ANY notice hereunder shall be sufficiently given to or served on the person to whom it is addressed if it is sent in a prepaid letter by the recorded delivery service (airmail if to an address outside the United Kingdom) addressed to that person at that person's last known address and shall be deemed to have been served three (3) days after the date of such posting

(1) For the purposes of this Agreement any notice shall be deemed to have been given to the personal representatives of a deceased Partner notwithstanding that no grant of representation has been made in respect of the deceased Partner's estate in England if the notice is addressed to the deceased Partner by name or to the

deceased Partner's personal representatives by title and is sent by prepaid letter by the recorded delivery service to the usual abode of the deceased Partner at the deceased Partner's death

(2) In this Agreement where the context permits the masculine gender shall include the feminine gender and the singular number shall include the plural number and vice versa

23. ANY disputes between the Partners (which expression shall where the circumstances so require include the representatives of a Partner) shall be referred in accordance with the provisions of the Arbitration Acts 1950 and 1979 and all statutory modifications or re-enactments thereof for the time being in force to a single arbitrator to be appointed by agreement between the Partners or in default of agreement by the President or one of the Vice Presidents of the time being of the Royal Institution of Chartered Surveyors

AS WITNESS the hands of the parties hereto the day and year first before written

Appendix XI

THE BUSINESS NAMES ACT 1985: SECTIONS 1–7

1. Persons subject to this Act

(1) This Act applies to any person who has a place of business in Great Britain and who carries on business in Great Britain under a name which –

(a) in the case of a partnership, does not consist of the surnames of all partners who are individuals and the corporate names of all partners who are bodies corporate without any addition other than an addition permitted by this Act;

(b) in the case of an individual, does not consist of his surname without any addition other than one so permitted;

(c) in the case of a company, being a company which is capable of being wound up under the Companies Act 1985, does not consist of its corporate name without any addition other than one so permitted.

(2) The following are permitted additions for the purposes of subsection (1) –

(a) in the case of a partnership, the forenames of additional partners or the initials of those forenames or, where two or more individual partners have the same surname, the addition of 's' at the end of that surname; or

(b) in the case of an individual, his forename or its initial;

(c) in any case, any addition merely indicating that the business is carried on in succession to a former owner of the business.

2. Prohibition of use of certain business names

(1) Subject to the following subsections, a person to whom this Act applies shall not, without the written approval of the Secretary of State, carry on business in Great Britain under a name which –

(a) would be likely to give the impression that the business is connected with Her Majesty's Government or with any local authority; or

(b) includes any word or expression for the time being specified in regulations made under this Act.

(2) Subsection (1) does not apply to the carrying on of a business by a person –

(a) to whom the business has been transferred on or after 26th February 1982; and

(b) who carries on the business under the name which was its lawful business name immediately before the transfer, during the period of 12 months beginning with the date of that transfer.

(3) Subsection (1) does not apply to the carrying on of a business by a person who –

(a) carried on that business immediately before 26th February 1982; and

(b) continues to carry it on under the name which immediately before that date was its lawful business name.

(4) A person who contravenes subsection (1) is guilty of an offence.

3. Words and expressions requiring Secretary of State's approval

(1) The Secretary of State may by regulations –

(a) specify words or expressions for the use of which as or as part of a business name his approval is required by section 2(1)(b); and

(b) in relation to any such word or expression, specify a Government department or other body as the relevant body for purposes of the following subsection.

(2) Where a person to whom this Act applies proposes to carry on a business under a name which is or includes any such word or expression, and a Government department or other body is specified under subsection (1)(b) in relation to that word or expression, that person shall –

(a) request (in writing) the relevant body to indicate whether (and if so why) it has any objections to the proposal; and

(b) submit to the Secretary of State a statement that such a request has been made and a copy of any response received from the relevant body.

4. Disclosure required of persons using business names

(1) A person to whom this Act applies shall –

(a) subject to subsection (3), state in legible characters on all business letters, written orders for goods or services to be supplied to the business, invoices and receipts issued in the course of the business and written demands for payments of debts arising in the course of the business –

(i) in the case of a partnership, the name of each partner,
(ii) in the case of an individual, his name,
(iii) in the case of a company, its corporate name, and
(iv) in relation to each person so named, an address in Great Britain at which service of any document relating in any way to the business will be effective; and

(b) In any premises where the business is carried on and to which the customers of the business or suppliers of any goods or services to the business have access, display in a prominent position so that it may easily be read by such customers or suppliers a notice containing such names and addresses.

(2) A person to whom this Act applies shall secure that the names and addresses required by subsection (1)(a) to be stated on his business letters, or which would have been so required but for the subsection next following, are immediately given by written notice to any person with whom anything is done or discussed in the course of the business and who asks for such names and addresses.

(3) Subsection 1(a) does not apply in relation to any document issued by a partnership of more than 20 persons which maintains as its principal place of business a list of the names of all the partners if -

(a) none of the names of the partners appears in the document otherwise than in the text or as a signatory; and
(b) the document states in legible characters the address of the partnership's principal place of business and that the list of the partners' names is open to inspection at that place.

(4) Where a partnership maintains a list of the partner's names for purposes of subsection (3), any person may inspect the list during office hours.

(5) The Secretary of State may by regulations require notices under subsection (1)(b) or (2) to be displayed or given in a specified form.

(6) A person who without reasonable excuse contravenes subsection (1) or (2), or any regulations made under subsection (5), is guilty of an offence.

(7) Where an inspection required by a person in accordance with subsection (4) is refused, any partner of the partnership concerned who without reasonable excuse refused that inspection, or permitted it to be refused, is guilty of an offence.

5. Civil remedies for breach of Section 4

(1) Any legal proceedings brought by a person to whom this Act applies to enforce a right arising out of a contract made in the course of a business in respect of which he was, at the time the contract was made, in breach of subsection (1) or (2) of section 4 shall be dismissed if the defendant (or, in Scotland, the defender) to the proceedings shows –
   (a) that he has a claim against the plaintiff (pursuer) arising out of that contract which he has been unable to pursue by reason of the latter's breach of section 4(1) or (2), or
   (b) that he has suffered some financial loss in connection with the contract by reason of the plaintiff's (pursuer's) breach of section 4(1) or (2).

   unless the court before which the proceedings are brought is satisfied that it is just and equitable to permit the proceedings to continue.

(2) This section is without prejudice to the right of any person to enforce such rights as he may have against another person in any proceedings brought by that person.

6. Regulations

(1) Regulations under this Act shall be made by statutory instrument and may contain such transitional provisions and savings as the Secretary of State thinks appropriate, and may make different provision for different cases or classes of case.

(2) In the case of regulations made under section 3, the statutory instrument containing them shall be laid before Parliament after the regulations are made and shall cease to have effect at the end of the period of 28 days beginning with the day on which they were made (but without prejudice to anything previously done by virtue of them or to the making of new regulations) unless during that period they are approved by a resolution of each House of Parliament.

   In reckoning this period of 28 days, no account is to be taken of any time during which Parliament is dissolved or prorogued, or during which both Houses are adjourned for more than 4 days.

(3) In the case of regulations made under section 4, the statutory instrument containing them is subject to annulment in pursuance of a resolution of either House of Parliament.

7. Offences

(1) Offences under this Act are punishable on summary conviction.

(2) A person guilty of an offence under this Act is liable to a fine not exceeding one fifth of the statutory maximum.

(3) If after a person has been convicted summarily of an offence under section 2 or 4(6) the original contravention is continued, he is liable on a second or subsequent summary conviction of the offence to a fine not exceeding one-fiftieth of the statutory maximum for each day on which the contravention is continued (instead of to the penalty which may be imposed on the first conviction of the offence).

(4) Where an offence under section 2 or 4(6) or (7) committed by a body corporate is proved to have been committed with the consent or connivance of, or to be attributable to any neglect on the part of, any director, manager, secretary or other similar officer of the body corporate, or any person who was purporting to act in any such capacity, he as well as the body corporate is guilty of the offence and liable to be proceeded against and punished accordingly.

(5) Where the affairs of a body corporate are managed by its members, subsection (4) applies in relation to the acts and defaults of a member in connection with his functions of management as if he were a director of the body corporate.

(6) For purposes of the following provisions of the Companies Act 1985 –

(a) section 731 (summary proceedings under the Companies Acts), and

(b) section 732(3) (legal professional privilege), this Act is to be treated as included in those Acts.

BYE-LAW 25(2) AND (3): DISCIPLINARY POWERS

25. (2) If in respect of a Member there is produced to the General Council:

(a) a certificate of conviction by a Court of competent jurisdiction for any criminal offence involving embezzlement, theft, corruption, fraud or dishonesty of any kind or any other criminal offence carrying on first conviction a maximum sentence of not less than twelve months imprisonment; or

(b) a notice in the London Gazette that he has been adjudicated bankrupt or a certified copy of a deed of arrangement he has entered into with or for the benefit of his creditors; or

(c) a notice in the London Gazette or the Edinburgh Gazette that a company which carries on practice as surveyors of which he is a director has had a winding-up order made in respect of it or has passed a resolution for voluntary winding-up (not being a member's voluntary winding-up); or

(d) other evidence satisfactory to the General Council that the Member or a company carrying on practice as surveyors of which the Member is a director has become insolvent; and

(e) a copy of a letter sent by the Institution to the Member
  (i) informing him of the powers of the General Council under this Bye-Law;
  (ii) giving him not less than 21 days notice of the date of the meeting of the General Council at which the matter is to be considered; and
  (iii) inviting him to make such submissions in writing as he may think fit;

the General Council after considering any submission as aforesaid may either:

(i) without further enquiry forthwith expel the Member from the Institution; or

(ii) refer the matter to the Disciplinary Board hereinafter mentioned for enquiry and action.

(3) The General Council may if they think fit temporarily suspend the Member from membership of the Institution pending such enquiry as last aforesaid.

BYE-LAW 26 AND 26A: DISCIPLINARY PROCEDURE

Bye Law 26

26. (1) The disciplinary powers of the Institution under paragraph (1) of Bye-Law 25 of this Section shall be exercised by a Professional Conduct Committee, a Disciplinary Board and an Appeals Board and sitting in private and respectively constituted and acting as follows.

(2) (a) At the first meeting of the General Council held after the Annual General Meeting the General Council shall appoint for the ensuing year a Professional Conduct Committee consisting of eighteen Members of the Institution of high standing including:

(i) not less than ten members of the General Council (excluding Associates); and

(ii) not less than one representative of each of the Divisions of the Institution, save in the cases of the General Practice Division and of the Quantity Surveyors Division which shall each have two representatives.

(b) If in the opinion of the Professional Conduct Committee the investigation of a particular complaint or allegation may require specialist knowledge or expertise not available to the Committee it may appoint a Member who in its opinion has such knowledge or expertise to serve adhoc as a full additional member of the Committee for the purpose of considering the particular complaint or allegation.

(3) The quorum of the Professional Conduct Committee shall be seven.

(3A) It shall be the responsibility of the President to appoint a panel (hereinafter referred to as 'the Permanent List') of not less than twenty-five Members of the Institution of high standing to serve when required on the Disciplinary Board and the Appeals Board pursuant to paragraphs (4) and (5) of this Bye-Law for a minimum period of three years.

(4) (a) The hearing of any charge referred to the Disciplinary Board under paragraph (6)(d) of this Bye-Law and of any matter referred by the General Council under paragraph (2) of Bye-Law 25 of this Section shall be by a panel consisting of not less than three nor more than seven members appointed by the President from the Permanent List and including at least one representative of the same Division of the Institution as that of the Member against whom the charge is preferred or in respect of whom the reference is made. The panel shall sit with a legally qualified assessor.

(b) If for any reason any member of the panel is during the course of the hearing unable to continue to attend the hearing, or any adjourned hearing, of any charge or matter, the remaining members of the panel, providing there are not less than two in number of whom at least one is a representative of the same Division as the Member concerned, may at their discretion proceed or continue with the hearing, but if the Member appears they shall do so only with his consent. In any case where the hearing is not proceeded with by the remaining members of the panel, and in any case where such remaining members do hear a charge or matter wholly or in part but are unable to arrive at any determination thereof the charge or matter shall be re-heard by a new panel.

(c) If at any time the President is of the opinion that it is for any reason impracticable for the hearing of any charge or matter to be completed by the panel appointed in that behalf, he shall direct that the case be re-heard by a new panel.

(d) Whenever a case is re-heard pursuant to sub paragraph (b) or (c) of this Bye-Law, any of the members of the original panel may be appointed to the new panel.

(5) (a) An appeals Board for the purposes of paragraph (8) of this Bye-Law shall consist of five or seven members appointed by the President from the Permanent List so that the number of members appointed shall be not less than the number of members of the Disciplinary Board who heard the case but not including any member of the Professional Conduct Committee or of the panel of the Disciplinary Board who was concerned in the investigation or hearing of the charge or matter in respect of which the appeal is made. An Appeals Board shall sit with a legally qualified assessor.

(b) If for any reason any member of the Appeals Board is during the course of the hearing unable to continue to attend the hearing, or the adjourned hearing, of an appeal, the remaining members of the Appeals Board, providing they are not less than three in number, may at their discretion proceed or continue with the hearing, but if the Member appears they shall do so only with his consent. Save as aforesaid the appeal shall be re-heard by a new Appeals Board.

(c) If at any time during the course of the hearing of an appeal, the Appeals Board is of the opinion that it is for any reason impracticable for it to complete the hearing, the President shall appoint a new Appeals Board to hear the appeal.

(d) Whenever an appeal is re-heard pursuant to paragraph (b) or (c) of this Bye-Law, any of the members of the original Appeals Board may be appointed to the new Appeals Board.

(6) (a) The Professional Conduct Committee shall investigate complaints or allegations made against a Member of unprofessional conduct or other breach of these Bye-Laws or Regulations by him. The Committee may sit with a legally qualified assessor.

(b) The Member shall be entitled to submit written observations or representations to the Committee on the subject-matter of the complaint or allegation and if in any particular case he desires to do so to appear before the Committee in person; and the Committee shall have power to require the Member to attend before them; to require him to produce any documents in his possession which it might consider to be relevant to its investigation; and to request the attendance of witnesses. If the Member fails to attend or otherwise to avail himself of his rights under this paragraph the Committee may proceed in his absence and without further reference to him.

(c) A Member appearing before the Committee may be represented by a solicitor or counsel and may, at the discretion of the Chairman, call witnesses on his behalf.

(d) The Committee may reject the complaint or allegation and acquit the Member but if in their opinion the complaint or allegation is justified they may (i) reprimand or severely reprimand the Member or, either separately from or in addition to the former, (ii) require from him such reasonable undertaking (whether oral or written) not to continue or repeat the conduct complained of or alleged as appears to the Committee to be desirable or (iii) refer the subject-matter of the complaint or allegation to the Disciplinary Board as a formal charge.

(7) (a) The Disciplinary Board (acting by a duly constituted panel thereof) shall consider all charges referred to it by the Professional Conduct Committee.

(b) The Member in respect of whom the charge is thus referred shall be entitled to appear before the Board either in person or by solicitor or counsel and to give evidence and call witnesses and cross examine witnesses called against him or adduce other evidence in his defence. If the Member fails to attend or otherwise avail himself of his rights under this paragraph the Board may proceed in his absence and without further reference to him.

(c) The Board shall have the power to require a Member appearing before it to produce any documents in his possession which it considers might be relevant to the charge or charges referred to it and to request the attendance of witnesses.

(d) The Board may dismiss the charge and acquit the Member or if they find the charge proved may take any of the courses of action listed in paragraph (1) of Bye-Law 25 of this Section either immediately or (if the Board think it appropriate to give the Member an opportunity to remedy his misconduct or other default) after an adjournment for such period not exceeding six months as the Board shall think fit.

(e) The Board shall likewise consider all matters referred to it by the General Council under paragraph (2) of Bye-Law 25 of this Section and after affording the Member an opportunity of being heard either in person or by solicitor or counsel and of adducing evidence on the nature and circumstances of the conviction, bankruptcy, composition or arrangement (as the case may be) may decide what if any penalty (within the limits of paragraph (1) of Bye-Law 25 of this Section) shall be imposed on him. If the Member fails to attend or otherwise avail himself of his rights under this paragraph the Board may proceed in his absence and without further reference to him.

(f) In the case of a Member who lives and practises his profession outside the United Kingdom any proceedings against him before the Disciplinary Board (including the constitution of the panel thereof) shall be subject to such modifications as shall for the time being be prescribed by the Regulations.

(g) If pursuant to the foregoing provisions of this paragraph the Board shall reprimand or severly reprimand or suspend or expel the Member he shall be entitled within such period (such period being not less than twenty-one days after the hearing in the case where the Member was present, and not less than twenty-one days after notification of the Board's decision by recorded delivery post to the Member's last address on the register of the Institution, in the case where the Member was not present) as may be prescribed by the Regulations to appeal against the penalty thus imposed on him but not against the finding of the Board. Any such appeal shall be by notice given or sent by registered post to the Secretary-General and shall specify the grounds to be relied on in support of the appeal.

(8) An Appeals Board shall consider any such appeal as aforesaid and after hearing the appellant or his solicitor or counsel (if any) may affirm or (within the limits of paragraph (1) of Bye-Law 25 of this Section) vary the penalty appealed against to one of greater or less severity, and may if they think fit make such order as the Board shall consider just for the payment by the Member of a sum of money in or towards payment of the costs incurred by the Institution in connection with the appeal or by the Institution to the Member in or towards his costs incurred in connection with the appeal.

(9) All decisions of the Disciplinary Board and of an Appeals Board shall take immediate effect and shall in due course be reported to the General Council and duly recorded; and the relevant Board may cause to be published in the Journal notice of the reprimand, severe reprimand, suspension or expulsion of a Member together with such particulars as the Board shall think desirable of the misconduct, breach of Bye-Laws, conviction or other matter for which the penalty in question was imposed. The Board may also notify such newspapers and other publications as it shall think fit of any such notice and particulars.

Provided that no penalty imposed by the Disciplinary Board shall take effect or be reported to the General Council or any notice thereof be published within the prescribed time for appealing therefrom or (if a notice of appeal is given) while the appeal is pending.

(10) If a Member is expelled his name shall be removed from the Register and he shall thereupon cease for all purposes to be a Member of the Institution. His diplomas of membership shall be immediately returnable and he shall not be entitled to use

any designation or description which implies membership or former membership of the Institution or of the Society or the Institute.

(11) (a) If a Member is suspended his diplomas of membership shall be immediately returnable and he shall not be entitled during the period of his suspension to exercise any of the rights or privileges of membership of the Institution or (in particular) to use any such designation or description as aforesaid. He shall however remain in all other respects subject to the provisions of these Bye-Laws and to the exercise of the Institution's disciplinary powers in respect of any contravention of those provisions committed by him during the period of his suspension.

(b) If a Member is suspended he shall forthwith notify his existing clients of the fact of his suspension.

(12) Once a Member has been notified that a complaint or allegation has been made against him or that a conviction, bankruptcy or other matter mentioned in paragraph (2) of Bye-Law 25 of this Section has been notified to the Institution he shall not be entitled to resign from membership of the Institution until all proceedings against him under this Bye-Law have been concluded; and any such proceedings may be continued notwithstanding his attempted resignation.

Bye-Law 26A

(1) Where the Professional Conduct Committee receive evidence that a Member has failed:

(a) to deliver to the Institution any certificate, report or other document required by the Bye-Laws and Regulations; or

(b) to comply with any provision of the Regulations made pursuant to Bye-Law 24(9A)

the Committee may, instead of dealing with him under the provisions of Bye-Law 26 or transmitting particulars of such failure to the General Council for the exercise of its powers under Bye-Law 31, as the case may be, adopt the following procedure in his case.

(2) The Committee may suspend such Member either forthwith or from such date as they may specify.

(3) (a) If the Committee decide to suspend a Member pursuant to paragraph (2) of this Bye-Law they shall forthwith notify the Member concerned and shall review their decision within twenty-eight days of the order for suspension taking effect.

(b) On such review the Member shall be entitled to make representations in writing for consideration by the Committee.

(4) (a) On a review the Committee shall, if the Member has remedied his default, revoke the order for suspension.

(b) If on the review the Member has not remedied his default the Committee shall resolve either to consider the matter under the provisions of Bye-Law 26 or to refer the matter to the Disciplinary Board as a formal charge and in either event shall continue the order for suspension and shall forthwith notify the Member in writing of their decision.

(c) Without prejudice to the Committee's powers under Bye-Law 26(6)(d) or to the powers of the Disciplinary Board under Bye-Law 26(7)(d), the Committee or the Disciplinary Board, as the case may be, shall revoke the order for suspension at any stage of their proceedings under Bye-Law 26 on receiving evidence that the Member has remedied his default. The duty to revoke the suspension shall rest with the Professional Conduct Committee until the date of any hearing of the Disciplinary Board, whereupon it shall rest with the Disciplinary Board.

(d) The Disciplinary Board shall, before taking any of the courses of action under Bye-Law 26(7)(d), revoke the order for suspension whether or not the Member has remedied his default.

(e) The Committee or Disciplinary Board as the case may be shall cause the facts of a Member's suspension and the revocation of any such order to be noted in the Member's Personal Record file.

(5) The consequences of suspension shall be the same as are set out in paragraph (11) of Bye-Law 26.

(6) Notice of a decision to suspend a Member under paragraph (3)(a) of this Bye-Law shall advise the Member of:

(i) the fact and effective date of the suspension; and

(ii) the date upon which the Committee will review the suspension; and

(iii) the Member's rights under paragraph 3(b) of this Bye-Law and the Committee's powers under paragraph (4) of this Bye-Law.

(7) Any Notice under this Bye-Law shall be sent by recorded delivery post to the Member's last address registered by the Institution and shall be deemed to have been effectively served on the Member forty-eight hours after the date of posting.

BYE-LAW 31: ERASURE FROM THE REGISTER

31. (1) If any member fails:
    (a) to pay any monies due from him to the Institution, whether in respect of his subscription, a levy or otherwise, within three calendar months from the date upon which they become due; or
    (b) if his subscription is payable by instalments in accordance with Regulations made pursuant to Bye-Law 28(1), to pay any instalment by the date on which it becomes due; or
    (c) to deliver to the Institution any certificate, report or other document required by any of these Bye-Laws or of the Regulations within three months from the date upon which it became due for delivery

    he shall automatically lose the right to attend or vote at General Meetings of the Institution and the General Council shall have the power to erase his name from the register, declare that he be no longer a Member and demand the surrender of his diploma. If the General Council exercise such power they shall forthwith notify the person concerned.

    (2) The General Council may in general or in any particular case where they think it desirable defer exercising the power prescribed in paragraph (1) of this Bye-Law until such later dates as they may determine.

PROFESSIONAL INDEMNITY COLLECTIVE POLICY

The Royal Institution of Chartered Surveyors Professional Indemnity Collective Policy

IN CONSIDERATION of the Assured named in the Schedule hereto having paid the premium set forth in the said Schedule to the Insurers who have hereinto subscribed their Names (hereinafter referred to as 'the Insurers')

THE INSURERS HEREBY SEVERALLY AGREE each for the proportion set against its name to indemnify the Assured (as defined herein) in accordance with the terms and conditions contained hereunder or endorsed hereon,

PROVIDED THAT:

1. the liability of the Insurers shall not exceed the limits of liability expressed in the said Schedule or such other limits of liability as may be substituted therefore by memorandum hereon or attached hereto signed by or on behalf of the Insurers,
2. the liability of each of the Insurers individually in respect of such loss shall be limited to the proportion set against its name.

SCHEDULE

Whereas the Assured, as defined herein, have made to Insurers a written proposal bearing the date stated in the Schedule containing particulars and statements which it is hereby agreed are the basis of this Policy and are to be considered as incorporated herein:

INSURING CLAUSES

Now We, the Insurers, to the extent and in the manner hereinafter provided, hereby agree:

1. To indemnify the Assured against any claim or claims first made against them during the period of insurance set forth in the Schedule in

respect of any Civil Liability whatsoever or whensoever arising (including liability for Claimants' costs) incurred in the course of any Professional Business carried on by or on behalf of the Assured.

2. To indemnify the Assured against any loss or losses which, during the period specified in the Schedule, they shall discover they have sustained by reason of any dishonest or fraudulent act or omission of any past or present Partner, Director or Employee of the Firm(s) named in the Schedule, provided always that:
   (a) No indemnity shall be afforded hereby to any person committing or condoning such dishonest or fraudulent act or omission and the sums payable under this Policy shall be only for the balance of Liability in excess of the amounts recovered from the dishonest or fraudulent person or persons or their estates or legal representatives.
   (b) The annual accounts have been prepared and/or certified by an independent Accountant or Auditor.

   In addition such loss or losses shall include Accountants' fees incurred as a direct result of such loss up to £10,000 or such amount as arranged and agreed by Insurers.

3. To indemnify the Assured against any costs and expenses of whatsoever nature incurred by the Assured in replacing or restoring documents (as defined herein) either the property of or entrusted to the firm(s) contained in the Schedule or in the custody of any person to or with whom such documents have been entrusted, lodged or deposited, having been discovered during the period specified in the Schedule to be damaged, destroyed, lost or mislaid and which after diligent search cannot be found.

4. To pay eighty per cent (80%) of all costs incurred by the Assured in connection with legal proceedings taken by the Assured for the recovery of Professional Fees due to them for professional work done, subject to the following conditions:
   (a) The Assured must advise Insurers immediately of their intention to institute such proceedings.
   (b) No claim shall attach unless Insurers have been advised by their legal advisers that such action could be pursued with the probability of success.
   (c) Insurers' liability under this section shall not exceed £5,000.

5. The liability of the Insurers shall not exceed for any one claim under this Policy the sum specified in item 4 of the Schedule except as provided for under Insuring Clause 4 but Insurers shall in addition indemnify the Assured in respect of all costs and expenses incurred with their written consent in the defence or settlement of any claim which falls to be dealt with under this Policy, provided that if a payment in excess of the amount of indemnity available under this

Policy has to be made to dispose of a claim, the Insurers' liability for such costs and expenses shall be of such proportion hereof as the amount of indemnity available under this Policy bears to the amount to dispose of that claim.

6. If an amount is specified under Item 5 of the Schedule this amount shall be borne by the Assured at their own risk, and Insurers' Liability shall only be in excess of this amount.
   The excess shall not be applicable to:
   (a) claims or losses falling under Insuring Clauses 3 and 4 of the Policy.
   (b) claims arising out of any libel or slander.
   (c) costs and expenses incurred with Insurers' consent.

SPECIAL INSTITUTION CONDITIONS

Insurers will not exercise their right to avoid this Policy where it is alleged that there has been non-disclosure or mis-representation or facts or untrue statements in the proposal form, provided always that the Assured shall establish to Insurers' satisfaction that such alleged non-disclosure, mis-representation or untrue statement, was free of any fraudulent intent.

However, in any case of a claim first made against the Assured during the period of this insurance where (1) they had previous knowledge of the circumstances which could give rise to such claim and (2) they should have notified the same under any preceding insurance, then, where the indemnity or cover under this Policy is greater or wider in scope than that to which the Assured would have been entitled under such preceding insurances (whether with other Insurers or not), Insurers shall only be liable to afford indemnity to such amount and extent as would have been afforded to the Assured by such preceding insurance.

Where the Assured's breach of or non-compliance with any condition of this Policy has resulted in prejudice to the handling or settlement of any claim, the indemnity afforded by this Policy in respect of such claim (including costs and expenses) shall be reduced to such a sum as in Insurers' opinion would have been payable by them in the absence of such prejudice.

In the event of any dispute or disagreement between the Assured and Insurers regarding the application of these Special Institution Conditions, such dispute or disagreement shall be referred by either party for arbitration to any person nominated by the President for the time being of The Royal Institution of Chartered Surveyors.

DEFINITIONS

1. 'PROFESSIONAL BUSINESS' is understood to apply to advice given or services performed of whatsoever nature by or on behalf of the

Firm or Firms named in the Schedule wherever or by whomsoever given or performed and shall extend to apply to any Assured whilst holding an individual appointment in respect of work incidental to the Assured's business, providing the fee (if any) with respect to such advice or services is taken into account in ascertaining the income of the Firm.

In the event of any dispute arising between the Assured and Insurers as to the correct interpretation of the definition of Professional Business the facts shall be submitted to the President for the time being of The Royal Institution of Chartered Surveyors or his nominee whose decision shall be binding on both parties.

2. 'THE ASSURED' shall mean:
   (a) those persons named under Question 4 of the last completed proposal form and any other person or persons who may have subsequently become Partner(s)/Director(s) in the Firm(s) during the period specified in the schedule.
   (b) any former Partner(s)/Director(s) of the Firm(s) for services performed for and on behalf of the Firm(s) including retired Partner(s)/Directors remaining as Consultants to the Firm(s).
   (c) any person who is or has been under a contract of service for and/or on behalf of the Firm(s).
   (d) the Estates and/or the legal representatives of any of the persons noted under (a), (b) or (c) hereof in the event of their death, incapacity, insolvency or bankruptcy.
   (e) any company (other than a Partnership) named in Item 1 of the Schedule.
3. 'FIRM(S)' Wherever the word 'Firm(s)' appears herein the same shall be deemed to read 'the Firm(s) named in the Schedule or the predecessors in business of the said Firm(s)'.
4. 'DOCUMENTS' shall mean deeds, wills, agreements, maps, plans, records, books, letters, Certificates, Computer System Records, forms and documents of whatsoever nature whether written, printed or reproduced by any other method (other than bearer bonds, coupons, bank notes, currency notes and negotiable instruments).
5. 'ANY CLAIM' or 'ANY LOSS' shall mean any one occurrence or all occurrences of a series, consequent upon or attributable to one source or original cause. This definition shall also mean the discovery of the dishonesty of any person(s) and shall constitute one occurrence or one original cause.

EXCLUSIONS

The Policy shall not indemnify the Assured against any claim or loss:

1. In respect of which the Assured are entitled to indemnity under any

other insurance(s) except in respect of any excess beyond the amount which would have been payable under such insurance had this Policy not been effected.

2. Arising out of any circumstances or occurrence which has been notified under any other Policy or Certificate of Insurance attaching prior to the inception of this Policy.
3. Arising out of any dispute between the Assured and any present or former employee or any person who has been offered employment with the Assured.
4. Arising out of death or bodily injury of an employee under a contract of service with the firm(s) whilst in the course of employment for or on behalf of the Assured.
5. Arising from or brought by a firm, company or organisation in whom any partner(s)/director(s) have a controlling interest unless such a claim or claims are brought against the Assured firm(s) by an independent third party source.
6. Arising from the use of any motor vehicles by the Assured in circumstances in which the provisions of the Road Traffic Acts apply.
7. Arising out of the ownership by the Assured of any buildings, premises or land or that part of any building leased, occupied or rented by the Assured.
8. In respect of dishonest or fraudulent acts or omissions committed by any person after discovery, in relation to that person, of reasonable cause for suspicion of fraud or dishonesty.
9. Arising out of or in connection with any trading losses or trading liabilities incurred by any business managed by or carried on by the Assured.
10. Arising from liability for any amount in respect of liquidated damages or penalties which attaches due to liability assumed by the Assured under contract or agreement which would not otherwise have attached in the absence of such contract or agreement.
11. Arising from survey/inspection and/or valuation reports of real property unless such surveys/inspections and/or valuations shall have been made:
    (a) by a Fellow or Professional Associate of the Royal Institution of Chartered Surveyors (RICS); or
    by a Fellow or Associate of the Incorporated Society of Valuers and Auctioneers (ISVA); or
    by a Fellow or Associate of Faculty of Architects and Surveyors (FFAS); or
    by a Fellow or Associate of Royal Institute of British Architects (RIBA); or
    by a Fellow or Associate of Royal Institute of Architects of Scotland (RIAS); or